Advances in Design and Testing of Future Smart Roads

Streets and roads constitute an enormous part of civil infrastructure and a large part of our cities – a social resource that must be properly managed and developed. Therefore, many road construction companies, contractors, transport and traffic administrations and municipalities are seeking new road design models that can withstand modern challenges and demands. *Advances in Design and Testing of Future Smart Roads: Considering Urbanization, Digitalization, Electrification and Climate Change* deals with adapting current road designs to better withstand these future challenges as well as optimizing their structural design. Furthermore, the book illustrates recommendations and models for street/road sections, including road sections with a reconfigurable design, which can be used in both reconstruction and new construction of roads.

Features:

- Covers road testbeds that meet the challenge of future urbanization, digitalization and electrification
- Provides recommendations for potential climate change impacts, including flooding and ice accumulation problems
- Introduces the concept of reconfigurable and removable streets including recommendations for corresponding street testbeds

This book will be of interest to road construction companies and contractors, transport and traffic administrations and municipalities, lecturers, researchers, students, and anyone interested in transport infrastructure and future road designs.

Advances in Design and Testing of Future Smart Roads

Considering Urbanization, Digitalization, Electrification and Climate Change

Dina K. Kuttah

CRC Press
Taylor & Francis Group
Boca Raton London New York

CRC Press is an imprint of the
Taylor & Francis Group, an **informa** business

First edition published 2023
by CRC Press
6000 Broken Sound Parkway NW, Suite 300, Boca Raton, FL 33487–2742

and by CRC Press
4 Park Square, Milton Park, Abingdon, Oxon, OX14 4RN

CRC Press is an imprint of Taylor & Francis Group, LLC

© 2023 Taylor & Francis Group, LLC

Library of Congress Cataloging-in-Publication Data
Names: Kuttah, Dina, author.
Title: Advances in design and testing of future smart roads : considering urbanization, digitalization, electrification and climate change / Dina Kuttah.
Description: First edition. | Boca Raton : CRC Press, 2023. | Includes bibliographical references and index.
Identifiers: LCCN 2022021407 (print) | LCCN 2022021408 (ebook) | ISBN 9781032248073 (hbk) | ISBN 9781032248103 (pbk) | ISBN 9781003280224 (ebk)
Subjects: LCSH: Intelligent transportation systems. | Infrastructure (Economics)
Classification: LCC TE228.3 .K88 2023 (print) | LCC TE228.3 (ebook) | DDC 625.7/94—dc23/eng/20220505
LC record available at https://lccn.loc.gov/2022021407
LC ebook record available at https://lccn.loc.gov/2022021408

ISBN: 978-1-032-24807-3 (hbk)
ISBN: 978-1-032-24810-3 (pbk)
ISBN: 978-1-003-28022-4 (ebk)

DOI: 10.1201/9781003280224

Typeset in Times New Roman
by Apex CoVantage, LLC

To my family and my teacher Prof. Sabah Said Razouki.

Contents

Preface

Parts of this book are originally from the work performed in Smart Streets project financed by Vinnova. The Smart Streets project is Sweden's largest research project about the streets of the future and runs from 2019 to the beginning of 2022. It is implemented by the Royal Institute of Technology (KTH), Chalmers University of Technology, VTI, Spacescape, Sweco, White, Edge, Norconsult, IVL, City of Stockholm, City of Gothenburg & Umeå municipality. The overall aim in producing this book is to extend knowledge on how to design and test the durability of roads and streets that meet the challenges of future urbanization, digitalization, electrification, and climate change.

The author would like to thank all project partners for their good collaboration during the course of the project. The author would also like to thank you who read this book. If you find something that you think can be done in a better way, do not hesitate to contact me.

Linköping
February 2022
Dina K. Kuttah

The author

Dina K. Kuttah received her BSc, MSc and PhD degrees in civil engineering from Al-Nahrain University, Baghdad, Iraq, in 1997, 2000 and 2004, respectively. In 1997, she was awarded the first national prize as outstanding student in civil engineering science in Iraq. From 2000 to 2008, she served as assistant lecturer and then lecturer at the same university. In 2008 she immigrated to Sweden with her family and started serving as an external lecturer at Uppsala University, department of Earth Sciences, as well as learning Swedish. From 2011, she joined the Swedish National Road and Transport Research Institute (VTI), where she is currently a senior researcher in pavement technology. Dina has led several research projects exploring the use of developed techniques in road construction materials testing and has been principal researcher of her research team in European projects related to road and transport engineering. Dina has published over 50 papers in peer reviewed journals and international conferences, and she has also organized series of workshops toward more knowledge exchange between Sweden and Japan in pavement engineering technology. Dina lives with her husband and two children in Linköping.

Acronyms and abbreviations

AADT	Annual Average Daily Traffic
AMA	(Allmän material – och arbetsbeskrivning), General material and working description
APT	Accelerated Pavement testing
AV	Autonomous Vehicles
BÖ/AG	Bitumen-bound base course
CBP	Concrete Block Pavement
CBÖ	Cement-bound base course with binder course
CRCP	Continuously Reinforced Concrete Pavements
FEM	Finite Element Method
HMA	Hot Mix Asphalt
HVS	Heavy Vehicle Simulator
JRCP	Jointed Reinforced Concrete Pavements
LCC	Life Cycle Cost
PJCP	Plain Jointed Concrete Pavements
QR code	Quick Response code
RUP	Removable Urban Pavements
SECTM	Structural Excavatable Cement-Treated Material
SEK	Swedish Crown
TRVK	The Swedish Transport Administration's technical requirements

1 Introduction

GENERAL STATEMENT

Infrastructure is the organizational structure necessary to operate a community; it can also be defined as the services and facilities needed for an economy to function. This term often refers to technical structures that support society, such as roads, bridges, water and sanitation resources, electrical grids, remote communications and so on.

By serving travelers, roads and streets unite people and foster economic growth (Norton, 2016). They sustain trade, and their development is necessary to shape the growth of the society that builds them.

Roads and streets are usually vulnerable to the effects of weather and traffic. More recently, new influencing factors have subjected road infrastructures to new challenges.

Therefore, it has become important to consider opportunities for innovative structural design of smart roads and streets that have flexible design and consider the effects of current and future challenges, namely, *urbanization, digitalization, electrification* and *climate change*. This will be achieved by assigning new functions to the roads that will influence their structural design. To reach this goal, this book starts by shedding lights on the importance and types of road networks; then a brief description of materials and typical structural designs of roads are presented. Since this book deals with possible adaptations of current roads to better withstand future challenges, the impact of each challenge on the structural design of roads has been explained. Furthermore, new structural designs of future smart roads have been suggested with possible instrumentation of testbeds for further assessments of the suggested technologies.

ROADS VS STREETS

It is worthwhile to highlight that the configuration of streets differs from that of roads from a geometrical and architectural point of view. A street normally has shops/stores or houses along both sides of it which facilitates public interaction. A road may also have buildings on either side though its main function is as a transportation route – a way of getting from one place to another, especially between towns. From the technical and structural point of view, as discussed in the current book, both roads and streets follow almost similar design standards and the structural design is controlled by traffic loading, construction material type and requirement, climate zones and other related factors. Therefore, the technical constructions given in the present book are applicable to both roads and streets. Correspondingly, both terms (streets and roads) have been used in this book when dealing with their structural design.

DOI: 10.1201/9781003280224-1

ROAD NETWORKS

A road network is a system of interconnected roads designed to accommodate wheeled vehicles and pedestrian traffic. The road hierarchy categorizes roads according to their functions and capacities. While sources differ on the exact nomenclature, the basic hierarchy comprises freeways, arterials, collectors and local roads.

In Sweden, the road network consists of 98 500 km of state roads, 42 300 km of municipal roads, 74 000 km of private roads with state grants, 16 600 bridges, 20 tunnels and 39 ferry routes. Of the state road network, 18 400 km are gravel roads (about 20% of the total road length) and 2 144 km are motorway, excluding roads link (Trafikverket, 2017).

ROAD MATERIALS AND STRUCTURAL DESIGN

In order to discuss the impact of future challenges on the technical design and construction of smart roads/streets, it is important to highlight first the most common materials used in streets and roads construction because the characteristics of the building materials control the overall structural design.

The design of road surfaces defines the combination of material layers that make up the road construction on which traffic runs. Generally, the structural design of the road consists of (1) surface course (concrete, asphalt, paver blocks or gravel) located at the top, which provides a surface that is safe and durable enough for the traffic anticipated to be using it, (2) a base layer (mixed gravel, sand, and filler) that creates a uniform and stable support for surface course, and (3) a subbase layer that forms the basis of the superstructure. Sometimes ground geotextiles are used and can be laid out, when the subgrade material is of such a nature that it needs to be separated from the subbase layer. Additionally, a frost protection layer is necessary in roads constructed in cold regions (see Figure 1.1).

Surface course

Base layer

Subbase layer

Geotextile
(if necessary)

Subgrade

FIGURE 1.1   Basic construction of a road superstructure.

ASPHALT ROADS

The first recorded use of asphalt as a road-building material was in Babylon (south of Baghdad, Iraq) in 625 BC. In AD 1595 Europeans exploring the New World discovered natural deposits of asphalt (Virginia Asphalt Association, 2020). Nowadays, asphalt mixtures are considered one of the major road construction materials used worldwide. The superstructures of modern asphalt roads consist usually of wearing course, binder course, unbound base layer, bound base layer, subbase, and possibly protective layers that are not needed in case of rock underlay. Figure 1.2 shows the design of different asphalt pavement structures built of different superstructures (TRVK Väg, 2011).

The thickness of the wearing and binder courses depends on many factors including wheel load and axle configuration, the contact pressure, the vehicle speed and load repetitions, and other factors as mentioned previously.

More about the design of asphalt road structures in Sweden can be found in TRVK Väg (2011).

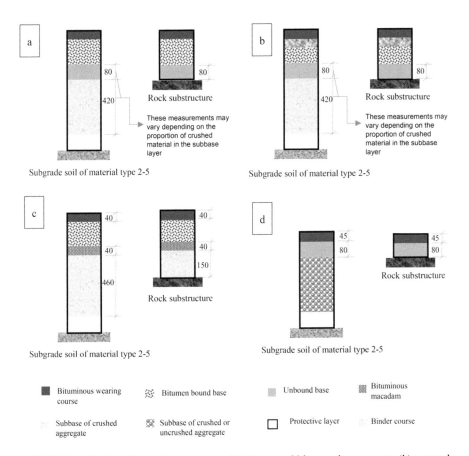

FIGURE 1.2   Design of asphalt pavement with (a) a gravel bitumen base course, (b) a gravel bitumen base course with binder layer, (c) a gravel bitumen base course with bituminous macadam and (d) a superstructure for the pedestrian and cycle path.

CONCRETE ROADS

The first concrete pavement in the world was built in Inverness, Scotland, in 1865. In 1891, the first US street was paved with concrete in Ohio (Concrete Construction Staff, 2014). Then the construction of concrete pavements spread to Europe in the 1920s (Williams, 1986). Concrete pavement design has over the years become a more important part of the promotion of concrete roads. There are in principle three different kinds of concrete road designs: plain jointed concrete pavements (PJCP), continuously reinforced concrete pavements (CRCP), and jointed reinforced concrete pavements (JRCP). The concrete roads in Sweden are PJCPs. Usually, the concrete roads consist of wear and base layers of cement concrete, cement-bound base, unbound base, subbase, protective layers and subgrade or rock substructure. Figure 1.3 shows the design of different concrete pavement structures according to TRVK Väg (2011).

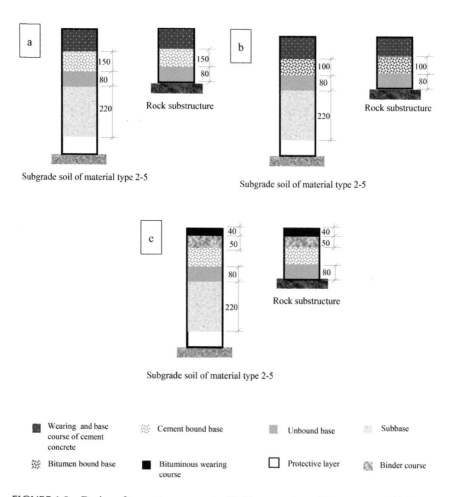

FIGURE 1.3   Design of concrete pavement with (a) cement-bound base course, (b) bitumen-bound base course, and (c) cement-bound base course with binder course.

PAVER BLOCK ROADS

The paver block roads are fairly close to the world's first paved roads in ancient Mesopotamia (Iraq), which were made of bricks and stone blocks (Lay et al., 2021). To date, many historic inner-city streets follow this category. Usage of concrete and stone blocks in paving is gaining popularity in areas where asphalt pavement does not last long. Paver block roads are used widely in municipality road application such as urban roads, airport taxiways, parking areas, ports and industrial areas. The performance of paver block roads depends upon block shape, size, thickness, type of bedding, joint sand and joint width. The laying pattern of blocks is also important, which affects the overall performance of the block pavement. The edge restraint is one of the features essential to stop mitigation of the block outward. The interlocking mechanism is one of the unique characteristics of the block pavements. The performance of block pavements largely depends on how well the interlock has been achieved (Ahmed and Singhi, 2013).

A typical superstructure of a block pavement according to the Svensk Markbetong (2019) consists of the coating layer of paver blocks, base and subbase layers resting on the subgrade. The subbase layer consists of unbound material or crushed concrete. The base layer may consist of unbound crushed material, bitumen bound gravel or cement-bound gravel. The coating consists of setting sand, joint filling sand and paver blocks. In Sweden, the materials of the unbound base and subbase layers must meet the requirements set out in AMA Anläggning (2020). The thickness of the unbound base and subbase layer depends on the material type, climate zone and traffic class as well as the choice of paver blocks thickness and material. Figure 1.4 shows examples of the construction of ground structure for traffic class 3, material type 4, climate zone 2 and frost hazard class 3 – see Svensk Markbetong (2019).

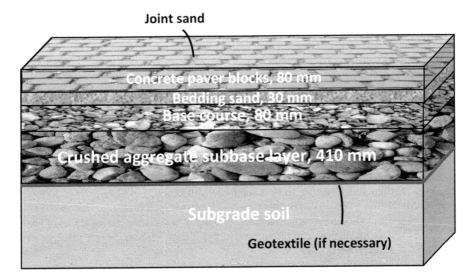

FIGURE 1.4 An example of a road section with paver block surface layer.

CHALLENGES THAT AFFECT THE STRUCTURAL DESIGN OF ROADS

All the aforementioned roads have different materials and structural design that varies based on many factors, including but not limited to, the type of the road, its place and hence the weather conditions, the design life of the road, the acceptable wheel load and vehicle length, the contact pressure, the vehicle speed and load repetitions. These influencing factors decide the structural design of roads/streets. Nevertheless, considering these influencing factors is no longer enough to guarantee that the current structural design can face the new challenges and serve the community as planned. To face the new challenges manifested by rapid *urbanization, digitalization, electrification* and *climate change*, the new structural designs of roads should be based on new guiding concepts and functions and be provided with new technologies that can withstand the new challenges. These functions may include, but are not limited to, the use of reinforced construction materials to withstand the platooning effect of autonomous vehicles, the conductive and inductive charging of electric vehicles while driving, and the new infraculvert system to reduce future roads excavation and maintenance work, in addition to flooding and snow control functions, and more.

All these new functions have different influences on the structural design of roads, therefore, different roadbed structures that meet the future requirements have been suggested in the following chapters.

Testing and instrumentation details of the new road designs have also been undertaken for further verification of the suggested technologies to withstand the influence of long-term traffic loading and climate conditions. Long-term durability tests of the suggested testbeds are recommended to optimize the social benefits before a broader use of the given solutions/technologies is adopted.

REFERENCES

Ahmed, A. and Singhi, B. (2013): "Overview on structural behaviour of concrete-block pavement," *International Journal of Scientific & Engineering Research*, 4 (7), 782–789. www.ijser.org/paper/Overview-on-Structural-behaviour-of-Concrete-Block-Pavement.html

AMA Anläggning. (2020): "Allmän material- och arbetsbeskrivning för anläggningsarbeten," *Svensk Byggtjänst*, ISBN 9789179170172.

Concrete Construction Staff. (2014): "Concrete pavement facts," *Infrastructure*. www.concreteconstruction.net/projects/infrastructure/concrete-pavement-facts_o

Lay, Maxwell, Metcalf, John and Sharp, Kieran. (2021): *Paving Our Ways: A History of the World's Roads and Pavements*, CRC press, Taylor and Francis group, Abingdon.

Norton, Peter. (2016): *Infrastructure: Streets, Roads, and Highways*, Oxford University Press, Oxford. https://doi.org/10.1093/acrefore/9780199329175.013.139

Svensk Markbetong. (2019): "Beläggning med plattor och marksten av betong - Projektanvisningar och rekommendationer" [Paving with slabs and paving stones of concrete -Project instructions and recommendations], *Tredje utgåvan, Svensk Markbetong*, Sweden, ISBN 978-91-519-3476-1.

Trafikverket. (2017): "Sveriges vägnät" [Swedish road network]. www.trafikverket.se/resa-och-trafik/vag/Sveriges-vagnat/ (Retrieved 2020–04–15).

TRVK Väg. (2011): "Trafikverkets tekniska krav Vägkonstruktion," TRV 2011:072, TDOK 2011:264, Trafikverket rapport, Borlänge.

Virginia Asphalt Association. (2020): "The history of Asphalt." https://vaasphalt.org/the-history-of-asphalt/

Williams, R.I.T. (1986): *Cement-Treated Pavements: Materials, Design and Construction*, Elsevier Applied Science Publishers Ltd., New York, NY.

2 Urbanization

GENERAL STATEMENT

Within the next 25–30 years, the urban population globally is expected to continue to grow, which means an increased use of urban streets and roads environments. Urbanization, combined with societal challenges and changing forms of life, also means more claims to the use of streets and roads, both different and new. This means that current regulations and standards for the design of streets and roads need to be changed and reassessed.

Contemporary developments in infrastructure show a link between investment, urbanization and economic growth (Li, 2017). As population and urban development has increased, the market has become an important role in financing infrastructure investments and helping to improve the efficiency of infrastructure investments. The most important infrastructure investments are usually related to electricity, water supply and sewage systems in addition to roads, railways and bridges.

Traditionally, electricity, water, fiber, sewage and heating pipes are buried into the ground. These subsurface infrastructure systems are now a basic requirement in any society. In Sweden, several of the networks used today were installed in the mid-20th century (Bergman and Olsson, 2017). This leads to increased maintenance work and increased maintenance costs for the companies that manage the networks. For example, the renewal of Swedish water and sewer lines involves an annual investment of SEK 1.9 billion, of which 50% goes to carry out maintenance operations that require traffic shutdown and excavation work (Bergman and Olsson, 2017). Therefore, new technologies in roads design should be considered to keep pace with the population growth, urban expansion and its burden on infrastructure. Keep in mind that selecting the technology should be based on its possible reduction of future maintenance and operation costs of the infrastructure services.

URBANIZATION AND ITS EFFECT ON THE STRUCTURAL DESIGN OF ROADS

Toward urbanization and improving the efficiency of infrastructure, new technologies related to infrastructure problems management have been adopted. The problems related to infrastructure of roads and streets have centered around high network maintenance costs – mainly those related to underground electricity, water, fiber, sewage and heating pipes maintenance. To avoid such high maintenance costs the ideas of using infraculvert technology has been started in Sweden as a form of new infrastructure investment that will lead to a reduction of the underground maintenance costs in the future. In 2017, an infraculverts system of about 1 800 meters long has been used in a relatively new district called Vallastaden in Linköping city for the first time. It involves systems for electrical, optical, water, sewage, waste suction and district heating, placed in a culvert. The culvert is the first of its kind and is made of

DOI: 10.1201/9781003280224-2

plastic with concrete chambers to connect the pipes and pull out the service lines. In other words, the infraculvert is a unified place for underground infrastructure; the long-term performance of the infraculvert under real traffic is of particular interest to be investigated. Technically, the infraculvert will provide a good environment and easy control for the pipe network, reduce the need for future excavation work, reduce disruption to society with fewer street closures during planned maintenance, and enable a vacuum waste collection system, which in turn leads to reduced heavy traffic (i.e., reducing the number of garbage trucks).

In the infraculvert used in Vallastaden, almost all infrastructure in the area is gathered. The innovative facility saved a lot of land that could be used for housing and parkland. According to Matton and Odenberg (2016), by using the infraculvert system, it is possible to build denser neighborhoods because not having to open shafts increases the buildable surface. The length of wires and cables can be reduced as they conventionally follow the street network, but in the culverts they can pass under buildings and through courtyards.

The infraculverts in Vallastaden were manufactured from 4 up to 32 meters and have an inner diameter of 2.2 meters. The diameter of the culvert depends on the number and size of the service pipes to be installed inside the culvert (Tekniska verken and Uponor, 2016). Based on visual in-situ inspection of the culvert along Uppfinnargränds gatan in Vallastaden, the distance between the upper edge of the culvert and the road surface was approximately 1 meter on the inspected section (Kuttah, 2019a) – see Figure 2.1. The infraculverts used in Vallastaden were 100%

FIGURE 2.1   Installation of infraculvert in Vallastaden-Linköping.

Source: Klas Gustafsson-Tekniska verken in Linköping, 2017

recyclable according to the manufacturer. This means that when, for example, a section needs to be replaced, it can be disconnected, lifted, recycled and replaced with a new section. In addition, the culverts can be used in different environments and can be built for any length to connect to concrete chambers that link together and carry the culvert pipe itself (Löwing, 2017).

Today there are empty places in the infraculvert installed in Vallastaden; in the future, these places can be used for new systems. Additionally, equipment for internal power and alarm equipment that warns of flooding, as well as smoke and gas development have also been installed.

The size of the pipe makes it easy to go in and draw even more cables if needed in the future. Figures 2.2 gives an example of how the pipes are located in the culverts.

FIGURE 2.2   Plumbing and sanitary system inside the infraculvert.

Source: Photograph by the author, 2019

In a test project, Sollentuna municipality installed 50 meters of infraculvert in an existing urban environment in the Växjö area. There, a traditional installation of underground service pipes had been initiated but proved expensive and time-consuming for the inhabitants of the municipality.

Nevertheless, there are also risks and disadvantages of the new infraculvert technology. Should a work-related accident occur in the confined environment of the infraculvert, relief work will be more difficult. In addition, the confined environment can lead to a leak in a system affecting other pipe networks, which leads to an increased risk of several systems being shut down at the same time compared with conventional installation of underground pipes system.

Even though from the engineering point of view the infraculvert is considered a sustainable solution that meet the challenges of future urbanization by improving the efficiency of infrastructures, deep studies are required to estimate the costs of using the new infraculvert technology as compared to the traditional pipes system. In this context, a study has been carried out by Al-kutubi and Yrlund (2020) to identify the economic feasibility comparing infraculverts and traditional pipe systems. The LCC analysis has been limited to the knowledge available in the current status. Therefore, the calculations have focused on excavation, construction and maintenance costs.

The results show that there is a tipping point of around 40 years where a clear difference can be seen over the profitability of the two systems. The infraculvert is theoretically cheaper when the simulation is carried out at an interval of (+/-) 20% of the total price. The study has also showed that the infraculvert has a lower maintenance cost, which is one of the driving costs in the simulation.

On the other hand, the culvert system will be placed on or at least pass under the city's streets and even the heavily trafficked roads. This means that the infraculverts will be subjected to dynamic stresses caused by passing cars and trucks in addition to the static stresses. Therefore, the embedding of an infraculvert within the roadbed and its interaction with the applied traffic loading should be evaluated thoroughly before a wider use of the system. Validation of the infraculvert with respect to its capability to withstand long-term traffic loading should be on focus. This action is very important, especially if the infraculverts are placed at shallow depths and become closer to the road surface for one reason or another.

RECOMMENDATIONS FOR TESTBEDS THAT MEET THE CHALLENGE OF FUTURE URBANIZATION

Though infraculverts have been used in local projects in Sweden, their long-term performance has not been determined under accumulated traffic loading, different ground and climate conditions. Correspondingly, this section suggests different testbeds that can be adopted for the long-term evaluation of infraculverts using accelerated loading testing. This test is important before a wider use of infraculverts is implemented in Sweden or any other country.

A testbed can be constructed and tested in a site that can provide an excavation depth of about four meters. This depth should allow the placement of the top edge of the infraculvert at least one meter's depth from the top of the final road surface. The testbed can be instrumented with various sensors and the traffic loading is applied by a heavy vehicle simulator (HVS) – see Note 2.1.[1] Usually, by using the HVS, the traffic loading can be simulated rapidly, and one can control and monitor the pattern of loading during testing.

In Vallastaden, the infraculverts have been placed under paver block streets. For this option, two testing alternatives are possible. Either the infraculvert in the testbed is placed along the loading direction as shown in Figure 2.3 or transvers to the loading direction as shown in Figure 2.4.

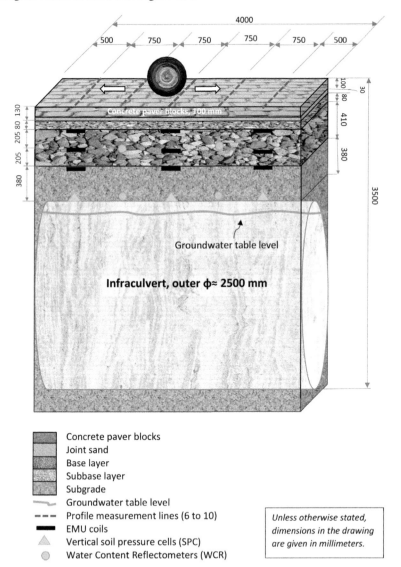

Unless otherwise stated, dimensions in the drawing are given in millimeters.

FIGURE 2.3 A recommended testbed constructed with infraculvert placed along the traffic direction. A road section with paver block surface, not to scale.

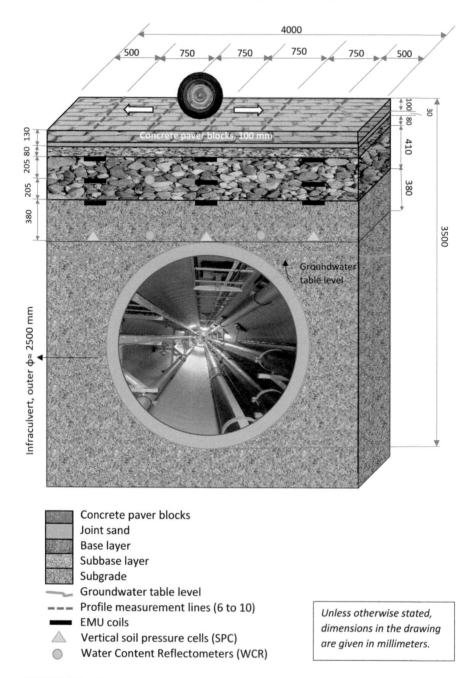

FIGURE 2.4 A recommended testbed constructed with infraculvert placed transverse to the traffic direction. A road section with paver block surface, not to scale.

The other option is to test the infraculverts underneath an asphalt road surface. A full-scale testbed of asphalt surfacing with an infraculvert underneath can be constructed according to the local technical standards for roads/streets constructions (e.g., in Sweden based on TRVK Väg, 2011).

Two alternative positions are also possible here; namely, the infraculvert is placed along the loading direction as given in Figure 2.5 or transvers to the loading direction, as illustrated in Figure 2.6.

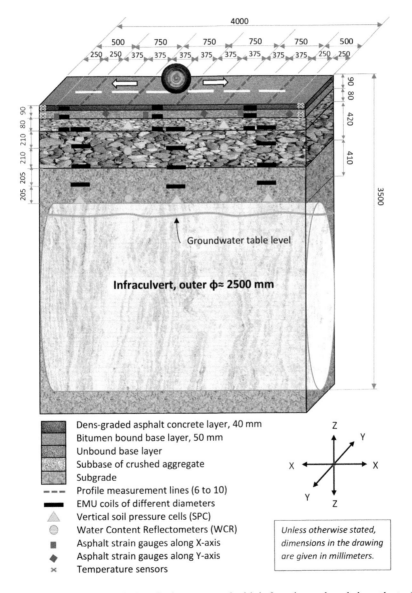

FIGURE 2.5   A recommended testbed constructed with infraculvert placed along the traffic direction. A road section with asphalt surface, not to scale.

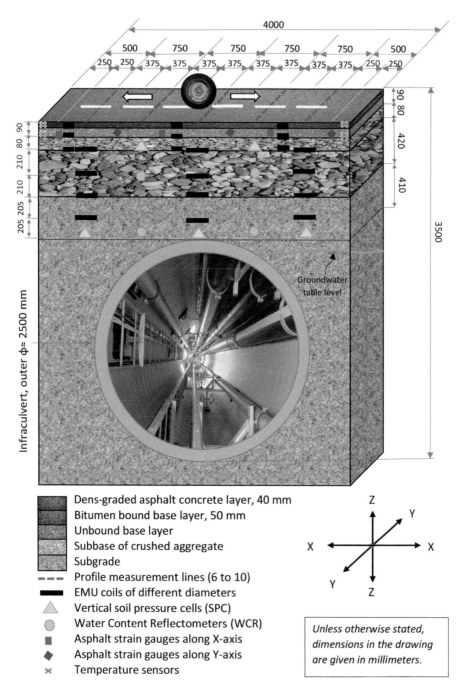

FIGURE 2.6 A recommended testbed constructed with infraculvert placed transverse to the traffic direction. A road section with asphalt surface, not to scale.

These testbeds will provide a huge database that correlates the sustainability of the infraculvert to the used building materials, adopted design, climate conditions and number of passes of equivalent standard axle (e.g., magnitude of traffic loading for the simulated number of years). The use of the collected data will be critical in making any decisions related to the future and wider use of infraculverts.

NOTE

1 Note 2.1 Accelerated testing of roads – https://www.vti.se/download/18.403857da1759746 17d27379/1592806490808/heavy-vehicle-simulator-hvs1.pdf

REFERENCES

Al-kutubi, A. and och Yrlund, J. (2020): "INFRASTRUKTURKULVERT – En jämförelse av ekonomisk lönsamhet mellan infrastrukturkulvert och traditionellt ledningssystem," Examensarbete, Örebro universitet, Institutionen för naturvetenskap och teknik. id: diva2:1447237.

Bergman, Filip and Olsson, Niklas. (2017): "Beräkningsverktyg till strategisk planering av framtidens ledningsbundna infrastruktur-Utveckling av modell för LCC- och LCA-analyser av ledningsbunden infrastruktur," M.Sc. thesis, Linköping University, Sweden.

Klas, G. (2017): "Projekt infrakulvert Vallastaden," Tekniska verken i Linköping. http://docplayer.se/15270154-Projekt-infrakulvert-vallastaden-klas-gustafsso-vvd-tekniska-verken-i-linkoping.html

Kuttah, Dina. (2019a): "Field visit and visual inspection of the infra culvert in Vallastaden on the 23 of October 2019 between 15:30–16:00 at Lärdomsgatan 14." Linköping.

Li, Z. (2017): "Infrastructure and urbanization in the people's Republic of China," ADBI Working Paper No. 632. Available at SSRN: https://ssrn.com/abstract=2898677 or http://dx.doi.org/10.2139/ssrn.2898677

Löwing, Joakim. (2017): "Framtidens lösning," Affärsstaden, Linköping, Sweden. https://affarsstaden.se/esb-article/framtidens-losning/

Matton, Maria and Odenberg, Camilla. (2016): "City planning to foster sustainable behaviour," M.Sc. thesis, International Business Administration Program, Linköping University, Sweden.

Tekniska verken and Uponor. (2016): "Infrakulvert Vallastaden 2016." Faktablad 75044 2016 12. https://issuu.com/uponor/docs/infrakulvert-vallastaden-uppslag/1

TRVK Väg. (2011): "Trafikverkets tekniska krav Vägkonstruktion," TRV 2011:072, TDOK 2011:264, Trafikverket rapport, Borlänge.

3 Digitalization

GENERAL STATEMENT

Since this book aims to develop models for future multifunctional roads that meet different challenges, the challenge of digitalization (e.g., autonomous vehicles) is one of the most demanding factors that may require new standards for roads/streets design.

Unfortunately, this aspect has received little attention so far. Therefore, the most important question on how autonomous vehicles (AVs) and other related digital technologies will affect the structural design of streets and roads structures will be discussed in this section.

DIGITALIZATION AND ITS EFFECT ON THE STRUCTURAL DESIGN OF ROADS

The introduction of autonomous vehicles on a broad scale is expected to create new requirements and standards for infrastructure design in order to enable their navigation and ensure the safety of other road users (Albino et al., 2015; Bösch et al., 2018).

In this context, a large number of studies have been carried out on the technologies required to operate autonomous vehicles on public roads, as well as analyses of the interaction and communication between autonomous vehicles and the road environment. However, there is limited research on how autonomous vehicles affect the design and performance of roads/streets structures. The introduction of autonomous vehicles into the fleet will mostly affect the surface and base layer, as these layers support the bulk of the stresses applied onto the pavement structure (Steyn and Maina, 2019).

The main impact of autonomous vehicles on the structural design of roads may be the effect of more channeled traffic load and platooning. Platoon driving involves groups of vehicles that can travel very closely, similar to a train; it is not limited to autonomous vehicles. Both light and heavy vehicles can be operated in platoons (Huggins et al., 2017). This can lead to higher traffic volumes of more uniform vehicle types and loads, which in turn will lead to a faster accumulation of pavement damage – something which has been confirmed experimentally according to Noorvand et al. (2017). In addition, platoon driving can lead to seasonal changes in pore water pressure in underlying layers of the pavements, thus weakening the subgrade soils by pumping (Saarenketo, 2018). With precision driving, the vehicles regularly travel in the same path within the lane over and over. This means that the load distribution and deterioration of road pavement will be different compared to current pavements. This fact will affect how pavements are designed and constructed for autonomous vehicles. Designing roads that can cater to AVs is important now, because whether the autonomous vehicles are here now or in 20 years, we are designing the roads for at least 30 years (see Figure 3.1).

DOI: 10.1201/9781003280224-3

FIGURE 3.1   One of Linköping's first autonomous minibuses runs along a two-kilometer route around Campus Valla.

Source: Photograph by the author, 2022

If a convoy of autonomous trucks are to be supported, roads maintenance programs also need to take into account increased loads from platoon driving.

Autonomous vehicles affect truck traffic, especially the choice of lane and positioning within the lane (Noorvand et al., 2017). Changes for existing infrastructure may result in an increased number and change in the location of lanes and load of the structure in ways for which it was not designed.

The use of autonomous and semi-autonomous transports can result in an even narrower use of the lanes (Litman, 2014; Lombardo, 2018; Snyder, 2018). According to Bowman (2016), the transition to autonomous vehicles assumes a concurrent reduction in lane width (about 25% reduction in freeway lane width) and an increase in roadway capacity (more than 50% according to several industry estimates). This assumption is consistent with what is given by Metz (2018).

Furthermore, the development of dynamic road markings makes it possible to optimize the use of existing square meters of roadways. By narrowing the lane width (along with speed limits), an additional lane can be added to the cross section. At all network levels, dynamic road markings bring automatic changes of travel direction in reach.

In conclusion, digitalization through the use of autonomous and semi-autonomous vehicles, dynamic road marking and intelligent merging supporting systems will result in increasing road network capacity. This will take place by means of dedicated lanes, temporary bypasses and flexible lane width that would result in additional loading of the pavement structure and hence a faster accumulation of pavement damage.

One of the aspects that needs to be looked at more closely is the elastic deformation bowl under truck tires that are following at close distances (due to platooning). De Beer (1992) demonstrated that axle loads applied at short intervals (due to closely spaced axles or at higher speeds) can cause the pavement to remain in a slightly deflected condition before the next axle causes another deflection bowl. The subsequent deflections are typically higher for subsequent loads, causing greater elastic deflection than if the load is applied at longer intervals that allow the pavement to rebound completely after each load application.

Monismith et al. (2000) evaluated the effect of channelized traffic (e.g., repetitive use of a single precise wheel path) and found that fatigue cracking failures could be accelerated by a factor of three or more by channelizing the traffic. With respect to pavement rutting, Harvey et al. (2000) evaluated the effect of channelized traffic (i.e., driving with zero lateral wander) using a Heavy Vehicle Simulator (HVS). The results demonstrated that the rutting on the channelized section is 25% to 45% greater than the rutting on the other section where the HVS moved laterally with a normal distribution. Steyn (2009) also studied the effect of wandering versus channelized traffic on the same test section. The experiments were carried out with a linear accelerated pavement testing (APT) equipment on sections paved with the same Hot Mix Asphalt (HMA) and having the same structural design. The results revealed that the permanent deformations caused by channelized trafficking is about 60% higher than the permanent deformations for the section loaded laterally with a normal distribution under the same testing conditions (speed and temperature). With reference to Noorvand et al. (2017), when trucks follow zero lateral wander, a faster accumulation of rutting evolution takes place. This finding is in line with the numerical studies that Erlingsson et al. (2012) have done, which found that a zero-wander state increased the rutting of flexible pavement structures by about 25%. Unfortunately, platoon driving could potentially create large cluster consisting only of heavy vehicles (Kulmala et al., 2019).

These observations suggest that a shift toward more channelized traffic from AVs can affect pavement deterioration, and that APT can be effectively utilized to quantify such effects.

According to Steyn and Maina (2019), introducing AVs can lead to more consistent driving conditions such as constant speed and fewer stops/starts actions at intersections thanks to better distributed traffic and steering at intersections. The cruising speed of the vehicle affects the contact time between the load and the bituminous road surface. A more consistent acceleration and deceleration of autonomous vehicles also leads to lower longitudinal stresses on the bituminous road surface, leading to less permanent deformation.

One of the recent studies on the flexible pavement design for autonomous and connected trucks is the one carried out by Al-Qadi and Gungor (2022). They have presented a framework that improves the analytical pavement damage accumulation approach by taking lateral position of loading as an explicit input in the pavement design.

Another technology related to the digitalization of future smart roads and streets is the controlling of vehicle speed. Achieving generally lower speeds in the city can create smoother flows and a better balance between modes of transport. In the long run this will lead to a reduction in separation, which means that one can save space. Saving space indicates also that one can give less space for car traffic and more space for shared and active mobility. As a result of these changes, it is possible to increase both the overall traffic capacity of the streets and create opportunities for more space for a range of functions that our cities need to meet the various challenges addressed in this book. However, achieving lower speeds requires more modern digitalized technologies able to control the vehicle speeds effectively.

One of the new speed control technologies is active speed bumps based on dynamic speed assurance solution. The "Actibumps" system works through radars which detect speeding vehicles and then activate a hatch on the road that lowers into the ground, creating an inverted speed bump to slow the car down. The system allows everyone who drives at the right speed to pass on a flat road and therefore does not pose a problem for bus drivers (Edeva, 2019).

Active speed bump systems can be considered a gradual transformation into more multifunctional and flexible streets/roads that can have a higher overall traffic capacity than before (including public transport passengers, cyclists, pedestrians and motorists).

This technology has been used in Sweden recently in a few counties (e.g., Uppsala, Linköping, Malmö and Skåne). In Linköping, the active speed bump was built on Djurgårdsgatan at the intersection with Väbelgatan. The street has a lane in each direction and annual average daily traffic, abbreviated (AADT) of 13 500 vehicles per day in 2015 (Börefelt and Nilsson, 2016). The street is served by bus traffic every 20 minutes and is an emergency route (for police, ambulance and fire brigade). See Figure 3.2.

Based on initial studies, it has been found that the average speed of free motor vehicles is lower with active speedbump technology than without, and usage of the system increases the traffic passability (Edeva, 2021). However, a higher proportion of motor vehicle drivers give preference to crossing unprotected road users when using the active speed bump system (Börefelt and Nilsson, 2016).

Other technologies related to the digitalization of the future streets and roads are the use of sensors embedded in the road surface that help increase road safety and allow vehicle-to-infrastructure communications. An important point is that the lifetime of the sensors needs to match that of the road surface or the infrastructure in which it is embedded, and an appropriate power supply must be provided. Not least, engineers need to manufacture sensors at a unit cost compatible with large-scale deployment. According to Clapaud (2017), this requires connected concrete, known in the industry as "functionalized concrete." This new technology enables sensors to be integrated into the surface of roads and streets via ducts and tunnels to monitor their physiochemical properties during their lifetime, thereby optimizing maintenance intervals. According to Clapaud (2017), temperature sensors are now available to help in managing salt-spreading operations in and around the city. Such sensors require cables to be laid in the road to provide electric power and gather data into a control point, and these constraints have limited the large-scale deployment

FIGURE 3.2 A Swedish active dynamic speed bump that only affects those who speed.

Source: Edeva

of these sensors. In Alterpave EU, a project embedding data collection sensors in the road body has been studied (Kuttah et al., 2018). The Smart Labelling Solution developed in the Alterpave project is a cheap solution that facilitates and enhances future road maintenance and rehabilitation, allowing tracking of the pavement components throughout their whole life. The solution includes a database for easy storage of the information of the road infrastructure. This information can be easily added and modified. Also, the information can be accessed directly via web page or with a mobile device. The QR codes used for accessing the information can be read with all QR scanners available in the market. The labels can be applied in outdoor conditions on several materials, including porous ones. However, further studies are required to further assess this technology and make it available for commercial use.

In the future, increasing use of sensors inside the pavement allows for more feedback from the infrastructure to the autonomous vehicle. Therefore, in the future it may be possible to limit heavy autonomous vehicles to certain lanes on streets and roads with several lanes in each direction. These lanes can then be designed with increased strength to be able to handle large volumes of heavy vehicles and closer distances (Johnson and Rowland, 2018).

Full-scale accelerating testing can be seen as a final design verification that checks the long-term performance of functionalized concrete and sensors. Nevertheless, it is too early at this stage to suggest a road testbed to verify the durability of these new technologies, as they have not yet been released for commercial use.

RECOMMENDATIONS FOR TESTBEDS THAT MEET THE CHALLENGE OF FUTURE DIGITALIZATION

From a design perspective, autonomous vehicles are a major factor in the digitalization challenge of the future, which will affect the design approaches and materials used for roads.

As mentioned previously, the narrow lateral wander pattern of autonomous vehicles can result in greater focus on rut-resistant pavements. Therefore, existing road and street design approaches and materials need further adaptations to withstand more effectively the expected increase in roads ruts and deformations if these roads are opened up to AVs. In this context, it is possible to strengthen the road where autonomous vehicles drive to better handle large volumes of heavy vehicles and platooning in several ways.

This can mean extra lanes constructed next to existing lanes, which can be more difficult to apply inside urban environments where buildings exist on both sides of the street. For infrastructure that has not yet been built or is in the planning phase, it is possible to consider how the current design might respond to any future requirement. While the ultimate requirements are unlikely to be clear, a design that considers how changes would be made in the future should be developed. The design could consider what type of strengthening actions could be easily undertaken in the future, how the pavement would be designed and constructed, and when it would be required within the design life. Other factors may be included in future studies, within the framework of existing road and street design standards; for example, the weight of autonomous trucks, acceleration rate and operational speed.

An interesting application that may become useful to resist increased rutting caused by autonomous vehicles is what is known as a "strip road" concept. A strip road consists of two narrow, parallel strips along the constructed pavement for the two wheel-tracks of vehicles. With the controlled wandering of autonomous vehicles, the "strip road" option could save large costs in the construction and maintenance

of the pavements. A reinforced strip road can address the higher incidence of channelized load application, where the design and construction of the wheel path has a
higher standard than between and outside the wheel path.

In order to enable the use of the test data to improve the understanding of the
expected autonomous vehicle effect on existing road pavement structures, this paragraph proposes three testbeds, that meet the challenge of future digitalization in
terms of using AV. Figure 3.3 shows a roadbed with strengthen strip (concrete or
steel mesh, e.g., 50 cm wide) to be built under the wheels of the autonomous vehicle.

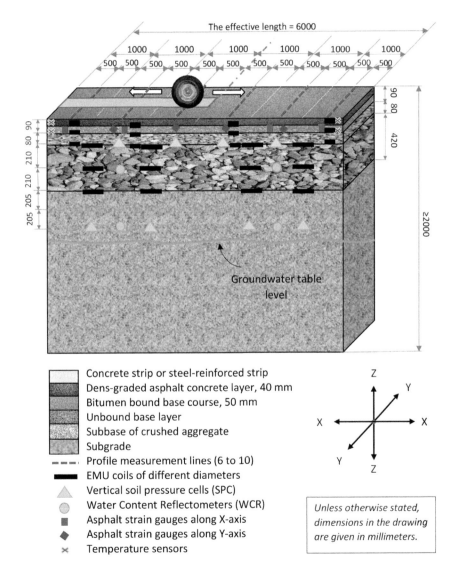

FIGURE 3.3   A recommended testbed constructed with concrete strip or steel-reinforced
strip. A road section with asphalt surface, not to scale.

Though the concept of combining asphalt and concrete in one lane to withstand AV traffic is something that has not been tried yet, combined pavements of concrete (in the high traffic volume lane) and asphalt (in the low traffic volume lane) have been used for many years with good results.

In Sweden, a short road segment on the E4 at Södertälje was constructed of concrete pavement in the right lane to serve heavy traffic and asphalt pavement in the left lane to serve light traffic. The concrete pavement lane was operated for 21 years without any maintenance measures being carried out. Rutting during this time was about 0.4 mm per year, but some kind of surface maintenance was needed. The joint between the lanes had lost sealer and was in need of sealing. The concrete section was instead overlaid with asphalt in 2009 during maintenance work on the adjacent bituminous section (Hultqvist and Dolk, 2015).

In Germany, combined paving with asphalt and unreinforced concrete has been used since the early 1980s and is still widely used in North Rhine-Westphalia. The road authority there thinks that it is a very economical construction method on road segments with a high traffic load and a large proportion of heavy vehicles. The same experience has also been adopted in South Africa between the capital Johannesburg and the port city of Durban and also in Oregon in the United States.

Proposals on how a combined road surface could be carried out in Sweden, with continuously reinforced concrete and with unreinforced concrete are discussed in by Hultqvist and Dolk (2015). Generally, for a pavement of combined asphalt and concrete materials experience shows the importance of a properly designed joint between the different materials and the importance of water drainage system of the pavement structure (Hultqvist and Dolk, 2015).

There are also other structural designs of roads capable of withstanding AV traffic without using the reinforced strip concept. This can be done by making a completely different cross section for the lanes or the whole road to be used by AV. In such designs, the whole road could be constructed with structural layers that are thicker than usual to withstand platoon driving with autonomous vehicles, as shown in Figure 3.4. Note that making thicker structural designs for individual lanes could be more costly than building the whole road with thicker structural layers. Regarding the suggested thicker road structure shown in Figure 3.4, the thickness of the bound layers can be decided based on practical adaptations.

The third proposal for the design of a road structure capable of withstanding AV traffic is based on using asphalt reinforcement fibers to stabilize the upper wearing course as shown in Figure 3.5. Aramid fibers (commonly used in bullet-proof vests, tire cords and heat protection) have been shown to reinforce asphalt, reducing rutting and cracking and increasing the asphalt dynamic modulus "stiffness" (ACF Environmental, 2020).

Research would be required to validate the modern construction and maintenance of such testbeds under long-term high traffic volume by means of accelerated pavement testing. This can be done by evaluating the design by testing a testbed consisting of two identical halves. One of the halves will be constructed based on the current standards and conventional materials while the other half will be constructed according to the new adopted design and materials.

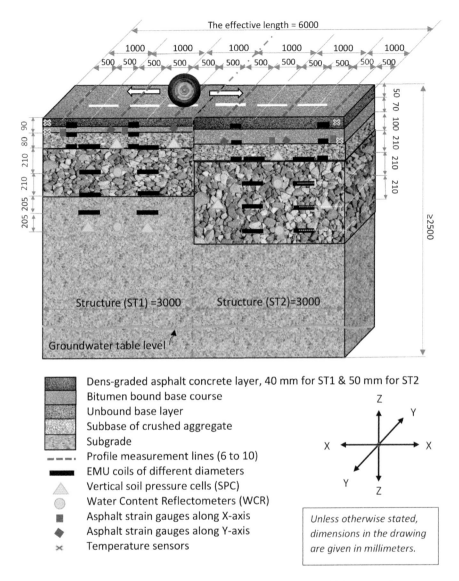

FIGURE 3.4 A recommended testbed constructed with different thicknesses of superstructure layers for flexible pavements, not to scale.

In case of accelerated pavement testing using the HVS on the testbeds proposed in Figures 3.3 to 3.5, the following points should be considered:

1. The lateral wander

Most HVS equipment has the function of setting the wander pattern used to apply the wheel load. The maximum width for load application determines the wander

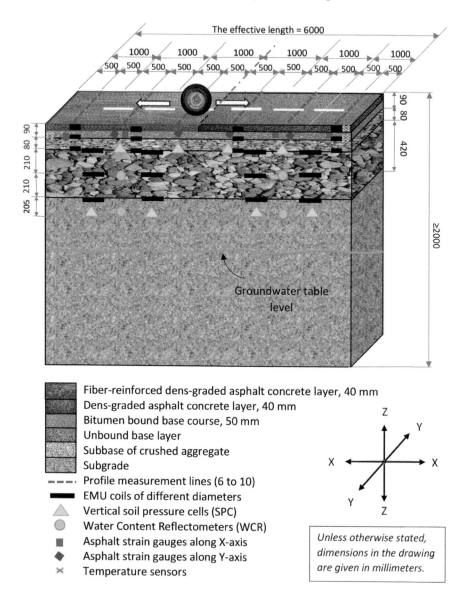

Fiber-reinforced dens-graded asphalt concrete layer, 40 mm
Dens-graded asphalt concrete layer, 40 mm
Bitumen bound base course, 50 mm
Unbound base layer
Subbase of crushed aggregate
Subgrade
- - - - Profile measurement lines (6 to 10)
▬▬▬ EMU coils of different diameters
△ Vertical soil pressure cells (SPC)
◯ Water Content Reflectometers (WCR)
■ Asphalt strain gauges along X-axis
◆ Asphalt strain gauges along Y-axis
✕ Temperature sensors

Unless otherwise stated, dimensions in the drawing are given in millimeters.

FIGURE 3.5 A recommended testbed constructed with standard and fiber-reinforced dens-graded asphalt concrete layer, not to scale.

pattern, and a transverse distance of about 50 mm is usually used. With the introduction of an autonomous vehicle, it should be possible to determine the wander pattern and expected speed of AVs to be used and recreate this in the HVS test or at least running a HVS test with zero lateral wander.

2. Loading mode during testing (i.e., unidirectional or bi-directional)

If loads are applied channeled (without lateral wander), this will cause significantly faster rutting development in unidirectional, rather than in bi-directional, loading mode (Tia et al., 2003). According to Snyder (2018), some streets in residential areas and those with low traffic volumes may have lanes that allow autonomous vehicles to drive in any direction needed. In this case, it is obvious to consider a bidirectional HVS test.

3. The speed of loads applications

According to Fagnant and Kockelman (2015), autonomous mobility is likely to lead to more stable speed profiles. Operating speeds during HVS tests are normally conducted at a standard speed. Although the speed range of most HVS devices are limited (typical operating range between 8 and 32 km/h), using speeds close to the actual speeds of the autonomous vehicles fleet is therefore fundamental in HVS test planning.

4. Controlled test temperature

It is well known that the temperature of bituminous surfacing materials is lower during the night as compared to during the day, resulting in higher stiffness of these materials and increased bearing capacity during the night. Therefore, it is desirable to have more driving at night with autonomous vehicles, based on their impact on the road structure. Studies conducted by Steyn and Denneman (2008) delt with the effect of pavement temperature (i.e., day versus night driving) on the permanent deformation of identical sections paved with HMA, showing about 790% higher permanent deformation at high temperatures than the low temperatures used in the study under channelized HVS loading. Increase in fatigue cracks due to loading in periods of low temperature have been reported extensively in the APT literature as highlighted by Steyn (2012). Correspondingly, the planning of the HVS tests should include a range of typical temperatures with pavement responses under different road conditions to develop a more general pavement responses model under expected field conditions. If the use of autonomous vehicles at night is expected, appropriate temperature programs should be selected to ensure that the HVS data reflects not only typical daytime bituminous layer response, but also expected nighttime responses.

5. Load magnitude

The load magnitude for AVs is expected to be higher than the usual load level limits. Therefore, the applied loads for HVS tests should be within the levels of those allowed for the autonomous vehicles including tire type and tire inflation pressure.

REFERENCES

ACF Environmental. (2020): "Asphalt reinforcement fibers." https://acfenvironmental. com/products/engineered-applications/roadway-reinforcement/innovative-aggregates-additives/asphalt-reinforcement/

Albino, V., Berardi, U. and Dangelico, R.M. (2015): "Smart cities: Definitions, dimensions, performance, and initiatives," *Journal of Urban Technology*, 22 (1), 3–21. http://dx.doi.org/10.1080/10630732.2014.942092

Al-Qadi, E. and Gungor, O. (2022): "Wander 2D: A flexible pavement design framework for autonomous and connected trucks," *International Journal of Pavement Engineering*, 23 (1), 121–136. https://doi.org/10.1080/10298436.2020.1735636

Börefelt, A. and Nilsson, A. (2016), "Utvärdering av Actibump i Linköping. Effekt på hastighet och väjningsbeteende." Serie nr 2016: 56, rapport 2016:56, version 1.0, Sverige.

Bösch, P.M., Becker, F., Becker, H. and Axhausen, K.W. (2018): "Cost-based analysis of autonomous mobility services," *Transport Policy*, 64, 76–91.

Bowman, J. (2016): "How autonomous vehicles will change the future of road design and construction," *FMI Quarterly*/September 2016. www.fminet.com/fmi-quarterly/article/2016/09/how-autonomous-vehicles-will-change-the-future-of-road-design-and-construction/

Clapaud, Alain. (2017): "Autonomous vehicles will require 'smart' roads," L'Atelier, published July 4, 2017. https://atelier.bnpparibas/en/smart-city/article/autonomous-vehicles-require-smart-roads

De Beer, M. (1992): "Developments in the failure criteria of South African mechanistic design procedure for asphalt pavements." In 7th International Conference on Asphalt Pavements, Nottingham.

Edeva. (2019): "30-säkring," Dokument nr 90 072-A, Sverige. cyclists_se.pdf (edeva.se).

Edeva. (2021): "Fallstudie Trafikverket," Document nummer 90 076, Version D. case_trafikverket_se.pdf (edeva.se).

Erlingsson, S., Said, S. and Mcgarvey, T. (2012): "Influence of heavy traffic lateral wander on pavement distribution." EPAM-4th Eur. Pavement Asset Manag. Conf. Statens väg-och Transp.

Fagnant, D.J. and Kockelman, K. (2015): "Preparing a nation for autonomous vehicles: Opportunities, barriers and policy recommendations," *Transportation Research Part A: Policy and Practice*, 77, 167–181. www.sciencedirect.com/science/article/pii/S0965856415000804?via%3Dihub

Harvey, J., Roesler, J., Coetzee, N. and Monismith, C. (2000): "CALTRANS accelerated pavement test (CAL/APT) Program." Summary Report: Six Year Period: 1994–2000. https://escholarship.org/content/qt5q7484fp/qt5q7484fp.pdf?t=qiv3md

Huggins, R., Topp, R., Gray, L., Piper, L., Jensen, B., Isaac, L., Polley, S., Benjamin, S. and Somers, A. (2017): "Assessment of key road operator actions to support automated vehicles," Austroads Publication No. AP-R543–17. https://wiscav.org/wp-content/uploads/2017/05/Austroads-Road_Agencies_Support_for_AVs.pdf

Hultqvist, B. and Dolk, E. (2015): "Betongbeläggning i tungt trafikerade körfält på motorväg- Exempel och erfarenheter från några olika länder," VTI rapport 837, Linköping, Sweden.

Johnson, B. and Rowland, M. (2018): "Automated and zero emission vehicles," Transport Engineering Advice. Infrastructure Victoria & ARUP. REP/261257, Issue 1 July 2018. 209 p.

Kulmala, R., Jääskeläinen, J. and Pakarinen, S. (2019): "The impact of automated transport on the role, operations and costs of road operators and authorities in Finland," EU-EIP Activity 4.2, Facilitating automated driving, Traficomin tutkimuksia ja selvityksiä 6/2019, ISBN (verkkojulkaisu) 978-952-311-306-0.

Kuttah, D., Indacoechea, I, Rodríguez, I., Lastra González, P., Blas, E., Casado, R., Boysen, R., Planche, J. and Trussardi, L. (2018): "ALTERPAVE Methodology," Deliverable 5.2, Alterpave European project.

Litman, T. (2014): "Autonomous vehicle implementation predictions: Implications for transport planning," *Transportation Research Board Annual Meeting*, 42, 36–42. https://doi.org/10.1613/jair.301

Lombardo, Jessica. (2018): "How will autonomous vehicles impact how we build roads-highways aren't going away, they're just going to change and fast," *Asphalt Contractor Magazine*, January 17, 2018. www.forconstructionpros.com/asphalt/article/20985026/how-will-autonomous-vehicles-impact-how-we-build-roads

Metz, D. (2018): "Developing policy for urban autonomous vehicles: Impact on congestion," *Urban Science*, 2, 33. www.mdpi.com/2413-8851/2/2/33

Monismith, C.L., Deacon, J.A. and Harvey, J.T. (2000): "WesTrack: Performance models for permanent deformation and fatigue." Pavement Resaerch Center, Univ. California, Berkeley.

Noorvand, H., Karnati, G. and Underwood, S. (2017): "Autonomous vehicles: An assessment of the implications of truck positioning on flexible pavement performance and design," *Transportation Research Record Journal of the Transportation Research Board*, 2640, 21–28. https://doi.org/10.3141/2640-03

Saarenketo, Timo. (2018): "Maintenance of infrastructure when automated driving takes over," CEO, Roadscanners Group, Aurora Summit, Olos, January 17, 2018. https://vayla.fi/documents/20485/421308/Timo+Saarenketo+Aurora+Pres+handout.pdf/e4dcbbbc-6146-48f8-a60f-4b9ea4dc1e78

Snyder, Ryan. (2018): "Street design implications of autonomous vehicles," *PUBLIC SQUAR, A CNU Journal, Transportation*. www.cnu.org/publicsquare/2018/03/12/street-design-implications-autonomous-vehicles

Steyn, W.J. (2009): "Evaluation of the effect of tire loads with different contact stress patterns on asphalt rutting." In GeoHunan Conference, Changsha.

Steyn, W.J. (2012): "Significant Findings from full-scale accelerated pavement testing," *NCHRP Synthesis*, 433. Transportation Research Board, Washington DC.

Steyn, W.J. and Denneman, E. (2008): "Simulation of temperature conditions on APT of HMA mixes." In 3rd International Conference on Accelerated Pavement Testing, Madrid.

Steyn, W.J. and Maina, J.W. (2019): "Guidelines for the use of accelerated pavement testing data in autonomous vehicle infrastructure research," *Journal of Traffic and Transportation Engineering* (English edition), 6 (3), 273–281.

Tia, M., Byron, T. and Choubane, B. (2003): "Assessing appropriate loading configuration in accelerated pavement testing." In 82nd Annual Meeting of the Transportation Research Board, Washington DC.

4 Electrification

GENERAL STATEMENT

Although the electrification of vehicles puts new demands on the electricity grid, as well as on the street environment with charging stations, there is considerable interest from governments and the motor industry in switching to clean low carbon transport.

The electrification of road transport began a long time ago with electrically powered public transport such as trolleybuses (Brunton, 1992), but being limited to specific routes made trolleybuses difficult to integrate with motor buses. Since then, much progress has been made in this sector as the transfer of the physical contact between the vehicle and power source has changed from overhead to the road surface. In addition, the development of onboard storage of energy, such as batteries, has helped in producing electricity for electric vehicles when needed.

In order to further develop electric vehicles, attention has been given to several vehicle charging technologies, such as providing fixed charging stations at rest stops, or even by providing external power charging along the pavement when the vehicle is in motion (i.e., dynamic charging of vehicles). The dynamic charging technologies are particularly interesting for roads and streets designers as they directly affect the structural design of roads and their long-term performance under different traffic loads and climatic conditions. In order to adapt roads to future electrification, these technologies will need further verification as discussed in detail in this chapter.

ELECTRIFICATION AND ITS EFFECT ON THE STRUCTURAL DESIGN OF ROADS

Electric roads give electric vehicles the opportunity to charge while driving (dynamic charging of electric vehicles). They can be built in different ways and with different technologies. In Sweden, three different dynamic charging concepts have been adopted up to date; namely, charging via overhead lines, conductive rail in the road or inductive coils below the road surface.

The main advantage of electric roads is that the vehicles do not need to have such large batteries. A smaller battery is lighter, cheaper and climate-smarter.

For electric roads with overhead lines, vehicles can be charged with electricity from an overhead power cable, as trams, to recharge while in motion. This would mean having a big pantograph on the roof of the vehicle. However, this technology is outside of the purpose of this book as it does not affect the structural design of the road.

Conductive charging requires physical contact between the electronic device's battery and the power supply (Ken, 2011) – that is, metal-to-metal contact between the charger and the electric vehicle is required, like a tramway track.

DOI: 10.1201/9781003280224-4

Outside Arlanda (in Sweden), the world's first electric road was opened where a car can charge its battery while driving. Along two kilometers of the route there are electric rail lines. A contact on the underside (i.e., an electric rail in the roadway similar to a tram track) allows the electric truck to charge its batteries while driving via an electric rail in the road – see Figure 4.1.

The charging distance is divided into shorter sections and the power is only switched on when a vehicle passes.

The aim of this technology is to investigate whether electrified roads can be a way of reducing emissions from truck traffic, which today accounts for almost 30% of road traffic's greenhouse gas emissions.

In Lund (in Sweden), a new electric road was inaugurated in 2020, which involves rails being placed in the middle of the existing road. This technique is based on a new version of electric rails, as shown in Figure 4.2. It includes a stretch of road of about one kilometer. It will be able to charge electric vehicles both while driving and stationary (Hoseini, 2019).

FIGURE 4.1   Electric rails are mounted in a road segment north of Stockholm.

Source: eRoad Arlanda

The other type of dynamic vehicle charging is inductive charging. Inductive charging (also known as cableless charging) is a kind of wireless charge that uses electromagnetic fields to transfer energy between two objects via electromagnetic induction without a charging rail (Chen, 2016; Åhman, 2018; Janzon, 2019).

The technology consists mainly of three parts. The first has to do with the road infrastructure and consists of copper coils covered with rubber, as shown in Figure 4.3. This part is usually embedded at a depth of eight cm in the road structure along the middle of the lane. The second part is a receiver located under the chassis of the vehicles. The communication system provides real-time communication with each vehicle. The third part is a power station; "an underground system that sends energy to the roads infrastructure."

The world's first wireless electric road for heavy trucks and buses has been in use since 2020 outside Visby (Gotland, Sweden), consisting of about 1.65 kilometers of dynamic wireless charging (Swartling, 2021).

FIGURE 4.2   Electric rails for vehicle charging.

Source: Photograph by the author

FIGURE 4.3   Inductive unit for vehicle charging made of copper coils covered by rubber coating.

Source: Photograph by the author

According to Åhman (2018), conductive charging has advantages over inductive charging. One of them is the cost. It is often said that it will cost about 10–20 times more to use inductive charging. In addition, inductive charging requires even surfaces so as not to have power losses.

The costs, or rather the additional costs of performing maintenance of an electric road compared to a normal road, have been analyzed and assessed based on experience. The operation and maintenance cost of the three types of electric roads constructed in Sweden differ significantly. The technology that clearly provides the smallest cost increases is the inductive technology (SEK 20 000/km, year) per direction of travel. For overhead lines and conductive rail in the road the additional cost is (SEK 122 000/km, year) and (SEK 165 000/km, year) per direction of travel, respectively (Palo et al., 2020).

Other types of electrification of roads are Solar Roads as the one constructed along Route 66 in the United States and Tourouvre-au-Perche in France (see Figure 4.4).

The problem with solar roads is that the panels are made of glass, which of course has completely different properties than asphalt. The glass panels use a kind of texture pattern to allow the cars to brake at about the same distance as on asphalt, but more tests are needed, especially on wet surfaces. Also, there are questions about the long-term durability of solar roads and what happens when the roads need to be snow-cleared (Söderholm, 2018).

According to Obminska (2019), the world's first solar road in France is so broken that it is not worth repairing. Therefore, it is not recommended that this technology be further investigated in its current stage; only if the technology is developed significantly in the future.

FIGURE 4.4   A new solar road in Tourouvre-au-Perche village in France.

Source: kumkum/Wikimedia Commons

RECOMMENDATIONS FOR TESTBEDS THAT MEET THE CHALLENGE OF FUTURE ELECTRIFICATION

The performance of electric rail and inductive charging technologies should be verified for long-term usage before wider adoption, especially regarding their ability to withstand deformations caused by high traffic volume. Furthermore, the interaction between these technologies and the road structure should be checked for long-term usage. The embedding of electric rails in asphalt, which has a different response to temperature deformations, should also be investigated in the long-term when large temperature variations are expected (diurnal and seasonal).

To validate the design life of an electrified road construction, Figure 4.5 shows a proposal for a standard testbed section with both conductive and inductive electric

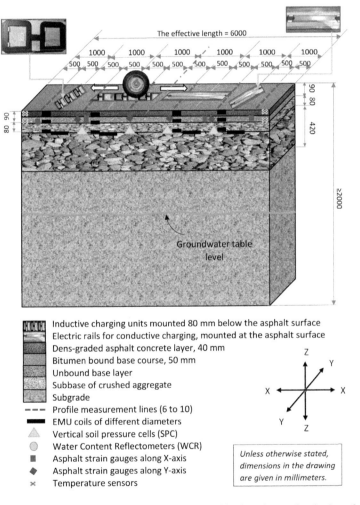

Inductive charging units mounted 80 mm below the asphalt surface
Electric rails for conductive charging, mounted at the asphalt surface
Dens-graded asphalt concrete layer, 40 mm
Bitumen bound base course, 50 mm
Unbound base layer
Subbase of crushed aggregate
Subgrade
--- Profile measurement lines (6 to 10)
▬ EMU coils of different diameters
△ Vertical soil pressure cells (SPC)
◉ Water Content Reflectometers (WCR)
■ Asphalt strain gauges along X-axis
◆ Asphalt strain gauges along Y-axis
× Temperature sensors

Unless otherwise stated, dimensions in the drawing are given in millimeters.

FIGURE 4.5 A recommended testbed constructed with charging technologies of electric vehicles, not to scale.

vehicle charging technologies that can be tested using the Heavy Traffic Simulator (HVS), with different groundwater levels.

In Figure 4.5, both technologies have been placed in two positions. One along the direction of traffic and the other at an angle of 45 degrees to the direction of traffic to simulate the passage of cars at an oblique angle, as in the case of a turn or an intersection.

The data to be collected from this test can assist in the development of the technologies, road designs and selection of road materials that best suit these technologies.

REFERENCES

Åhman, M. (2018): "Gotland först i världen med induktiv väg som laddar elbilen," Recharge October 10, 2018. www.mestmotor.se/recharge/artiklar/nyheter/20181010/gotland-forst-i-varlden-med-induktiv-vag-som-laddar-elbilen/

Brunton, L. (1992): "The trolleybus story," *IEE Review*, 38 (2), 57–61.

Chen, F. (2016): "Sustainable implementation of electrified roads: Structural and material analyses," Ph.D. thesis, KTH Royal Institute of Technology, Engineering Sciences, Department of Civil and Architectural Engineering, Stockholm, Sweden.

Hoseini, M. (2019): "VTI med i test av elvägar i Lund," VTI nyheter Publicerat April 15, 2019. www.vti.se/sv/nyheter/vti-med-i-test-av-elvagar-i-lund/

Janzon, Eva. (2019): "Israeler bygger elväg på Gotland," Världen idag, Nyheter·Publicerad 20:00, 2 maj 2019. www.varldenidag.se/nyheter/israeler-bygger-elvag-pa-gotland/repsdx!QAtcy5KSAm@NUasTOR8ymg/

Ken, N. (2011): "Wireless charging: Inductive or conductive?" www.phonearena.com/news/Wireless-charging-inductive-or-conductive_id16813

Obminska, Ania (2019): "Världens första solcellsväg – ett fiasko," Ny Teknik, Publiserad 2019–08–16. www.nyteknik.se/premium/varldens-forsta-solcellsvag-ett-fiasko-6968334

Palo, K., Eriksson, D., Söderberg, U. and Starkbeck, A. (2020): "Vägunderhåll och kostnader för olika typer av Elvägar," Trafikverket rapport TRV 2020/77969, Borlänge, ISBN: 978-91-7725-711-0.

Söderholm, Erik. (2018): "Solcellsvägar kanske aldrig blir verklighet – men ingen anledning att håna tekniken," Recharge, Publicerad 2016–07–07 08:11, uppdaterad October 1, 2018 11:26. www.mestmotor.se/recharge/artiklar/nyheter/20160707/solcellsvagar-kanske-aldrig-blir-verklighet-men-ingen-ide-att-hana-tekniken/

Swartling, C. (2021): "Gotland får väg med trådlös laddning," Transportnet, Gotland får väg med trådlös laddning – Transportnet.

5 Climate change

GENERAL STATEMENT

The global mean temperature for 2020 (January to October) was 1.2 ± 0.1 °C above the 1850–1900 baseline and 2020 is likely to be one of the three warmest years on record globally (State of the Global Climate, 2020).

Regarding the sea level rise, on average, since early 1993, the altimetry-based global mean rate of sea level rise amounts to 3.3 ± 0.3 mm/yr. The rate has also increased over that time. A greater loss of ice mass from the ice sheets is the main cause of the accelerated rise in the global mean sea level (World Climate Research Programme, 2018).

Based on the State of the Global Climate (2020), although understanding broad-scale changes in the climate is important, the most acute impacts of weather and climate are often felt during extreme meteorological events. These phenomena are manifested by heavy rain and snow, droughts, heatwaves, cold waves, and storms, including tropical storms, and can lead to or exacerbate other high-impact events such as flooding, landslides, wildfires and avalanches.

Climate change can affect the intensity and frequency of precipitation. Warmer oceans increase the amount of water that evaporates into the air. When more moisture-laden air moves over land or converges into a storm system, it can produce more intense precipitation; heavier rain and snowstorms (USGCRP, 2017).

In the United States, across most of the country, record heavy downpours are happening more frequently, delivering a deluge in place of what would have been routine heavy rain (Climate Central, 2015).

In Africa, very extensive flooding occurred over large parts in 2020, while India had one of its two wettest monsoon seasons since 1994, with nationally averaged rainfall for June to September 9% above the long-term average. It was also a very wet summer monsoon season over the Korean Peninsula, with the Republic of Korea experiencing its third-wettest summer on record, and parts of western Japan were affected by significant flooding in July. During the high rain fall in China for the period June to July 2020, at least 279 deaths were reported and the economic losses exceeded USD $15 billion (State of the Global Climate, 2020).

The UK is undergoing climate change with increased rainfall, sunshine and temperatures. The year 2020 was the third warmest, fifth wettest and eighth sunniest on record according to Kendon et al. (2021).

According to the Swedish National Knowledge Centre for Climate Change Adaptation at SMHI (2021), climate change is leading to precipitation in Sweden generally increasing, mostly in the north and during the winter months. However, there is great variation across the country and between the seasons. Climate calculations suggest that extreme precipitation will increase. In the future, flooding may become more common along the country's southern coastlines as a consequence of

DOI: 10.1201/9781003280224-5

rising sea levels. The impact on lakes and watercourses varies across the country; some areas are expected to experience greater flood risk and others lower risk (see Figure 5.1).

With respect to global snowfall, North America's most significant snowstorm of the 2019–20 winter occurred on 17–18 January in Newfoundland. St. John's received 75 cm of snow. A damaging ice storm in Oklahoma City saw power outages that lasted for days across more than half the city, and on 25 October 2020 occurred the earliest autumn date on which temperatures had fallen below −30 °C at a climate station anywhere in the United States (excluding Alaska) (State of the Global Climate, 2020).

With climate change, the duration of snow cover is expected to reduce. The extremely wet and warm winter of 2020 in northern Europe resulted in exceptionally low snow cover in many places – Helsinki experienced a record low number of snow-covered days, breaking the previous record by a wide margin (State of the Global Climate, 2020).

In the southern parts of Sweden, long-lasting snow cover will probably become rare (The Swedish National Knowledge Centre for Climate Change Adaptation at SMHI, 2021).

In summary, the primary climatic factors affecting road networks are precipitation, high water flows, ice, temperatures, sea levels and wind. In this context, the impact of climate change on the structural design of roads and streets will be discussed in the following paragraphs.

FIGURE 5.1   Major flooding in central Gävle (August 2021).

Source: Photograph by Mattias Gustavsson

CLIMATE CHANGE AND ITS EFFECT ON THE STRUCTURAL DESIGN OF ROADS

This paragraph discusses the effects of climate change on the structural design of roads and streets, especially for asphalt coated roads. In Sweden, the material requirements given in TRVK Väg (2011) should be considered. This includes requirements regarding the grain size distribution, organic content, frost danger, finished layers and safety during use. In addition, there are several points that should be taken into account during the technical design of roads to meet the challenges of future climate change.

The consequences of climate change on the road network are considerable. Increasing rainfall and greater flows lead to flooding, road flushing, an increased risk of collapse and landslide erosion. With rising temperatures, damage is shifted from being frost related to heat and water related. Precipitation affects the road superstructure mostly through the building up of groundwater and run-off in watercourses immediately after rain or due to snowmelt. Rising groundwater levels will increase pore pressure in the soil, weakening the natural slope stability of roads. High flows in large and medium-sized watercourses increase the risk of erosion that affect the slopes of watercourses. Heavy rains lead to high flows in small streams, mainly during summer and autumn, with the risk of erosion, flooding and washed-out roads and impacts on, for example, culverts (The Swedish Commission on Climate and Vulnerability, 2007) – see Figure 5.2.

Ground frosts and moderate and high temperatures play a role in the bearing capacity and durability of roads. Climate change will affect the future usage of AVs. According to Sohrweide (2018), autonomous vehicle technology does not work so well in snow, fog or heavy rain due to interference with vehicle sensors. Extreme temperatures are another challenge.

FIGURE 5.2 Half the way was gone (Umeå, November 2020).

Source: Photograph by CeGe Lillieroth

In the period between 1994 and 2001, there were around 200 events involving major damage due to high water flows in Sweden. Also, several high road slopes were washed away in Hagfors in 2004 after heavy rain. The total cost exceeded SEK 200 million. In the summer of 2006, a roadside slope was washed away after heavy rain and subsequent high flows. The road was repaired after two weeks at a cost of SEK 6 million (The Swedish Commission on Climate and Vulnerability, 2007).

Increased temperatures and reduced frost penetration will lead to different consequences for pavements. A shorter frost period means reduced deformation in the paving and subgrade but may require greater maintenance. Higher temperatures and groundwater levels mean increased rut formation.

When the subsurface temperature drops, the moisture in the unbound layers freezes to ice that binds the aggregate particles together. Frost penetration leads to an increased strength and stiffness of the unbound layers and subsoil. The process that forms ice also draws moisture into the freezing zone. When the frost thaws in the spring, the moisture in the soil increases, which can lead to weakened support for the pavement structure (Selezneva et al., 2008).

As the process of designing pavements moves toward mechanistic-empirical techniques, knowledge of seasonal changes in the pavement structural characteristics becomes critical. Specifically, information on frost penetration is necessary to determine the effect of freezing and thawing on the structural response of pavements.

Concrete constructions are more sensitive to salt and repeated freeze-thaw cycles. The number of zero passes – the number of days when the temperature passes the freezing point – is important for the road network and winter road maintenance (The Swedish Commission on Climate and Vulnerability, 2007). In summary, temperature changes affect pavement wear, rut formation, deformation and concrete repairs.

Willway et al. (2008) describe good examples of how local authorities can reduce the impact of climate change on pavements. Measures may include periodic monitoring of groundwater levels; changes in asphalt standards; long-term programs to locate and evaluate the efficiency and conditions of existing drainage in several municipalities; and programs to improve drainage, such as changing the aggregate used to one less prone to stripping and trialing of reinforcement of roads to reduce subsidence, as in Lincolnshire.

By expanding these activities and introducing more preventive, rather than reactive, maintenance measures, local authorities can succeed in adapting roadways to future climate changes (Willway et al., 2008).

Therefore, intensive work should be underway by the transport administrations and municipalities to develop specific requirements for road and street construction to be more resilient to climate change. The following section discusses new measures that should be considered while constructing smart roads capable of withstanding climate change.

RECOMMENDATIONS FOR TESTBEDS THAT MEET THE CHALLENGE OF FUTURE CLIMATE CHANGE

Ongoing climate change creates challenges for a developing society. The densification of cities increases surface runoff, reduces natural catchment areas and green areas, and increases stressors on stormwater systems. A combination of traditional road construction technology with climate change would risk aggravating the situation.

The various measures that can be taken to build roads capable of withstanding future climate change are, among others, those equipped with water reservoirs, new drainage systems, road heating systems and roads stabilized by different stabilizing technologies.

As highlighted in previous chapters, the surfacing layer (the wearing course) of a road can be made of asphalt, concrete, or paver blocks. Other types of materials can be used as surfacing layers in roads construction, but they do not fall within the focus of the testbeds present in the current section.

Regarding streets with paver blocks surfaces, new recommendations on the design of these streets are available now. These new designs take into account climate change by addressing floods and precipitation effects. Three paver blocks street systems have been defined by the Svensk Markbetong handbok (2019), namely system I, II and III based on drainage and stormwater storage systems. Systems I and II handle stormwater via infiltration through the surface layer of porous paving stones while system III consists of a bitumen-bonded structure with a top layer of dense paving stones where stormwater is led into the structure via wells or another type of side intake. Both systems I and II are divided into three subgroups (full infiltration, partial infiltration, no infiltration) depending on how much water is filtered out to the existing subgrade or led to the stormwater system (or other recipient). These streets systems are designed to reduce the amount, and improve the quality, of stormwater inside the street superstructure itself. For full infiltration systems, the rain infiltrates through the drained surface and further down through the superstructure to eventually reach the subgrade soil. Delay and storage of rainwater take place in the construction and to some extent in the underlying soil. In partial infiltration systems, drainage pipes are inserted into the subbase layer. In this way, residual water, after delay and partial infiltration into the subgrade, can be led on to the municipal stormwater network or adjacent delay facilities. In constructions with no infiltrations, all water that infiltrates through the paving stone is retained in a watertight reservoir (subbase) formed by an impermeable geotextile enclosing the structure. This construction is suitable when the subgrade has very low infiltration capacity (hydraulic conductivity), low bearing capacity, or when infiltration of surface water from the road area is not desirable.

For System III (bitumen-bonded construction with dense paving stones), full or partial infiltration of stormwater to the subsoil is not recommended. This is because during intensive precipitation, high flows can occur at wells and intakes, which means that the stormwater cannot be distributed evenly to the subgrade.

Figure 5.3 shows a design proposal for a paver block street of system I (according to the Svensk Markbetong handbook, 2019) that can better withstand weather changes such as flooding compared to the traditional system shown previously in Figure 1.4.

Note that the design shown in Figure 5.3 follows the same conditions as the one shown in Figure 1.4 for traffic class 3, material type 4, climate zone 2 and frost hazard class 3.

Joint sand

FIGURE 5.3   An example of a road section with a thick paver block superstructure to manage stormwater, system 1, not to scale.

AMA Anläggning (2020) introduces two new codes that specify material requirements for base and subbase in a superstructure that has a draining surface layer and where unbound layers should delay and drain the stormwater. All materials must be declared according to SS-EN, 13242. The unbound material for base layers should be in grade 2/32 (i.e., particle size between 2 mm and 32 mm) and for subbase layers in grade 2/90 (i.e., particle size between 2 mm and 90 mm).

It is important that the amount of material passing through the 2 mm sieve is limited to a maximum of 5% to provide high permeability and space for storing stormwater. For bearing capacity reasons, the amount of material between 2 and 4 mm is important, and as a guide value at least 10% should pass 4 mm sieve, according to Svensk Markbetong handbok (2019). Note that the type of paver blocks used in the surface layer in the new design (given in Figure 5.3) are drained, while the paver blocks used in the traditional design (given in Figure 1.4) are undrained (dense). Additionally, in traditional construction all the used construction materials – namely, the joint sand, bedding sand, base and subbase – contain higher amounts of material fraction, between 0- and 2-mm. Furthermore, the thickness of the subbase layer has been increased to 513 mm in the new climate adapted construction, as shown in Figure 5.3. This gives a total increase in the thickness of the entire street superstructure of 103 mm (as compared to the traditional one given in Figure 1.4). This extra thickness will serve as an additional reservoir for stormwater. Many of the new paver blocks street designs given in Svensk Markbetong handbok (2019) have been tested in full scale – see also Hellman (2017).

A new testbed is proposed in this section where porous asphalt has been used as a surfacing layer (instead of paver blocks); this design should be tested with respect to long-term performance at different traffic load and difficult climatic conditions, and compared with a similar structure with a dense asphalt surfacing. The new, proposed testbed retains the main advantage of extra reservoir space as seen in Figure 5.3, but with a porous asphalt surface layer as given in Figure 5.4. This figure shows how the proposed design takes climate change into account by managing flooding and stormwater via a thicker subbase as compared to a traditional standard one. The added thickness of the subbase layer will work as an additional reservoir to store the

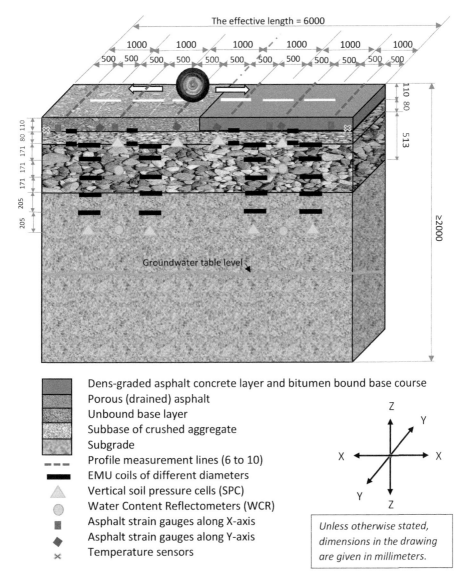

FIGURE 5.4 A recommended testbed constructed with stormwater control using porous asphalt and thick subbase layer, not to scale.

stormwater infiltrated to the road superstructure via the porous asphalt layer. Note that using porous asphalt may result in a weaker bearing capacity as compared to the traditional one since porous asphalt is usually more susceptible to deformation than undrained (dense) asphalt. Therefore, it is important to maintain a balance between the climate adapted design with drained asphalt and the higher bearing capacity of undrained asphalt. A long-term testing of the testbed given in Figure 5.4 will enable a good assessment of the deformations and bearing capacity of the two structures and hence a better understanding of their performance.

Furthermore, the road design given in Figure 5.4 can be combined with other functions that take into account electrification and digitalization challenges. Such possibilities are discussed in detail in Chapter 7.

Recently, many stormwater solutions have been developed to meet today's ambitions with stormwater management. Swales, rain gardens, rainbeds, storm basins, and smart traps are examples of interesting stormwater solutions.

Swale ditches transport water like pipes, but are designed as shallow, open, planted channels to convey runoff and remove contaminants. They are an alternative to a piped drainage system where space and inclination are available. Swales allow a slow flow of water and trap sediment to improve water quality (Global Street Design Guide, 2016).

Rain gardens have a special soil filter material that can remove pollution from road runoff. They are also called bioretention systems, flat bioswales, flow-through planters, or pervious strips. Some are designed to allow water to infiltrate down to underlying soils while others are designed to collect the treated water and transport the clean water downstream (Global Street Design Guide, 2016).

Rainbeds are new, compact stormwater systems that collect, delay and purify stormwater near the source. They combine green environments with stormwater management. Rainbeds are usually used on undrained surfaces and can be installed buried or above ground (see Figure 5.5).

According to CFS (2016), a storm basin is designed to capture and retain stormwater and capture contaminants such as sediment, debris, vegetation, nutrients,

FIGURE 5.5   A rainbed in Lidköping.

Source: Lidköping municipality, Vatten-Avlopp. Photograph by Adrian Lavén

coliform bacteria, oil/fat and dissolved metals (e.g., lead, copper, cadmium and chromium). Other stormwater solutions may include but are not limited to StormSack and StormSok as specified by CFS (2016).

As regards storage-tank stormwater systems, these tanks provide underground storage of stormwater. After a rain shower fills the storage tank, stormwater can flow into the sewage system, infiltrate into the ground or be reused. The system can be efficient and space-saving. The storage tank is a modular system that can be assembled to different heights. However, there are many different sizes and installation restrictions depending on the manufacturer's recommendations for specific traffic load and location conditions. Stormwater systems with storage tanks can be placed beneath a variety of surfaces including parking spaces, streets and access roads, and others, depending on the design requirements recommended by the manufacturer. Storage tank system technology is of particular interest for road design as it influences the structural design of roads and long-term performance under different traffic loading and climate conditions. If this technology is to be used in any country, a combination of local construction materials, local design standards and the manufacturer's installation recommendations should be considered. Figure 5.6 proposes a testbed with underground storage tanks for long-term performance evaluation under simulated traffic loading and climatic conditions. Note that the given structures are based on the Swedish standards and the dimensions should be adjusted based on the desired traffic loading, the selected storage tanks model and the construction advice of the manufacturers. The long-term performance of the final testbed should be evaluated under different surface flow and water table conditions.

Another relatively new solution to managing stormwater within the road body is the use of a Smart Trap. This solution has been developed and tested over four years by the University of Minnesota in the United States (see Figure 5.7). These tests have focused on two parameters: how effective the solution is in collecting sediment and how well it prevents collected sediment from swirling along at high water flows. Through a perforated partition inside the sand trap well, the water is prevented from swirling and dragging both old and new sediment. The Smart Trap system is a relatively new stormwater management technology that is interesting for road designers and requires long-term performance verification.

All the previously mentioned underground stormwater management technologies may be appropriate to provide instance management of large amounts of stormwater, which can prevent/reduce possible flooding of streets. Although swales and rain gardens bring a greener feel to the streets, they require space on the street that could be used for other services such as bike and electric scooter parking. Therefore, underground stormwater managing systems are recommended, especially for narrow streets – keeping in mind that the long-term performance of any system should be verified adequately before using. To give a greener feel to streets when using underground stormwater managing systems, vertical vegetation concepts with hanging plants can be used to make the narrow streets greener and more environmentally friendly (see Figure 5.8). Increasing green surfaces via vertical vegetated walls will help in reducing the impact of rainwater on drainage systems through the reduction of drained volume and peak flow.

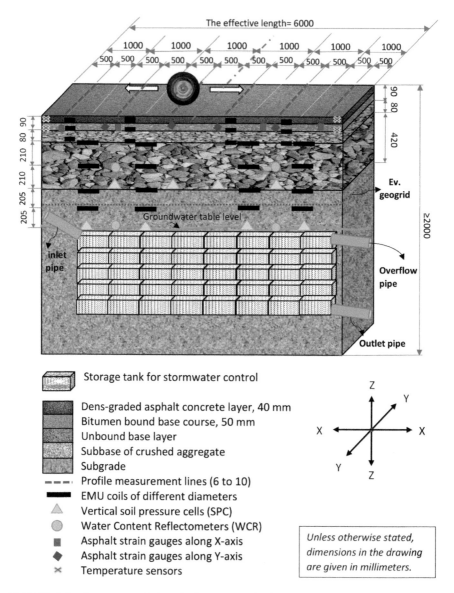

FIGURE 5.6 A recommended testbed constructed with stormwater control using storage tanks, not to scale.

Regarding snowfall: to avoid accumulation of snow and ice on the road surface and make roads anti-skid for current vehicles and future autonomous vehicles, it is recommended to embed heating systems in the road structure. By heating the streets or roads, it will be possible to achieve good results in increased accessibility and road safety during winter (as in the case of a road section built in Göteborgsbacken on RV40 in Jönköping).

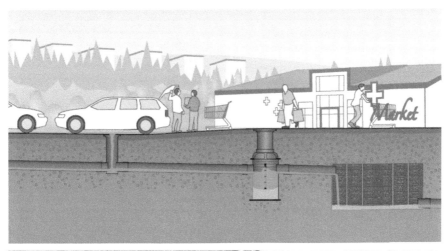

FIGURE 5.7 Smart Trap.

Source: Uponor Infra

According to Sundberg (2014), in Jönköping at Göteborgsbacken, heating pipes (coils) were laid on a distance of 1 500 meters. The heating coils are 2.5 cm in diameter and they are located under the asphalt, about 12 cm below the surface (see Figure 5.9). The loops are located approximately 25 cm apart and are connected to the district heating network in Jönköping Municipality. Since the pipes are laid under the top layer of asphalt, if the road needs to be repaired, the piping system may remain.

FIGURE 5.8   A green wall with hanging plants in Linköping.

Source: Photograph by the author, 2021

FIGURE 5.9   Pavement section with heating coils.

Source: Uponor, 2017

According to Sundberg and Lidén (2014), different types of energy can be used to heat the road, especially close to the city, such as district heating.

The evaluation of this system is that it has worked well and that savings for the operation of winter road maintenance are approximately SEK 150 000 in the form of fewer point efforts and savings in salting. There will also be gains in the form of accessibility and reduced traffic stops (Kåhre, 2020).

The operating costs for the heated road section can vary between SEK 300 000–600 000 per year (Sundberg, 2014).

Sundberg and Lidén (2014) add that it is not technically difficult to heat the road, but it should be done in an environmentally friendly and energy-efficient way, and at a reasonable cost. Therefore, many studies have been undertaken to evaluate road heating systems from an economic perspective, taking into account various factors involved. More details on investment and maintenance costs for specific road heating systems used in Sweden are provided by Sundberg (2014) and Kåhre (2020).

In Sweden, much detailed study has been conducted regarding the storage of solar energy, as it is the most universal technology, it is environmentally friendly and has low sensitivity to future energy price increases. The road heating system consists of a piping system embedded in the street or road that circulates stored solar energy so that snow and ice melt. In summer, the street/road acts as a solar collector and the energy is stored in the ground. Various benefits of the road heating system with stored solar energy have been reported by Sundberg and Lidén (2014). The system is energy-efficient, with up to 90% reduction in the energy consumed by free solar energy, and its efficiency in de-icing of the road in winter has been approved. Thereby, it helps reduce accident risk and improve traffic flow in winter. In other words, the heating system enables "smart roads" that prevent snow accumulation during winter and also reduce rut formation in the road via cooling during summer.

The pipe heating system is of particular interest for roads engineers as it directly affects the structural design of roads and streets and therefore it is highly recommended to verify long-term performance of various heating systems under different traffic loads and climatic conditions. According to Sundberg and Lidén (2014), the technical design and materials in the ground heating systems need to be examined and dimensioned more closely. Examples include the collection of water, insulation, the strength and temperature requirements of pipe materials, design that can withstand traffic loads, heat conduction in asphalt, design in a way so that maintenance is not complicated, and so on. Correspondingly, Figure 5.10 shows a suggested testbed for a road structure equipped with a pipe heating system for possible verification using an accelerated test facility.

As mentioned previously, several steps should be taken into account in the design of roads capable of withstanding the future challenges of climate change, including changing the standard asphalt, changing aggregate types to others less prone to stripping, and trialing of reinforcement of roads to reduce subsidence.

Therefore, strengthening and reinforcing options should be considered in street/road design, in order to achieve a stronger and more sustainable street/road that meets future challenges with climate change – mainly damage from soil moisture. This is also in line with the need for reinforcement to withstand automatization challenge in terms of autonomous vehicles.

Additives (such as limestone, cement or other) have long been used to stabilize subgrade soil and make the road superstructure less susceptible to damage due to

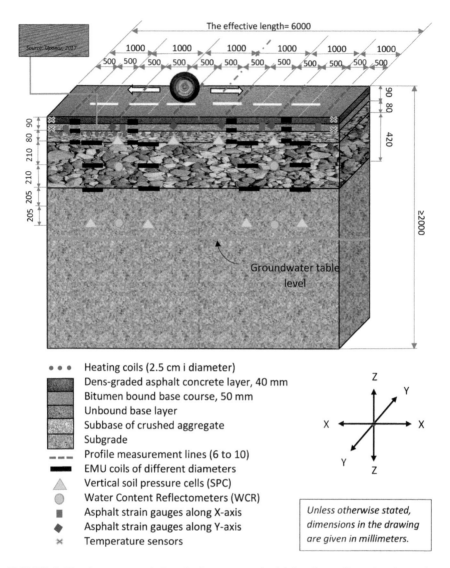

FIGURE 5.10 A recommended testbed constructed with heating coils under the asphalt surface, not to scale.

high traffic volume and the effects of water. Recently, new and environmentally friendly stabilization agents have been introduced to the market, including but not limited to forest rest products, acrylic-based stabilizers and enzyme-based stabilizers. Each stabilizer works optimally under certain conditions related mainly to the type of construction material to be treated, especially the amount of clay contents, the compaction moisture content and the curing period. For example, subgrade soil stabilized with enzyme-based stabilizers requires at least one month to maintain a noticeable stabilization effect. Note that good mix design protocol and reliable construction practices for each stabilization method should be adopted.

According to Franzén et al. (2012), surface stabilization usually takes place at the upper 350–400 mm of the subgrade, in order to obtain a satisfactory compaction.

The subgrade soil stabilization is of particular interest from the roads design and construction engineers' point of view as it helps strengthen the road super-structure to better withstand moisture damage, which is one of the biggest climate-related challenges. In this context, the long-term performance of various stabilized subgrade soils requires further investigation with simulated long-term traffic loads and different climatic conditions, and the resulting performance must be compared with the performance of an unstabilized road structure. Figure 5.11 illustrates a

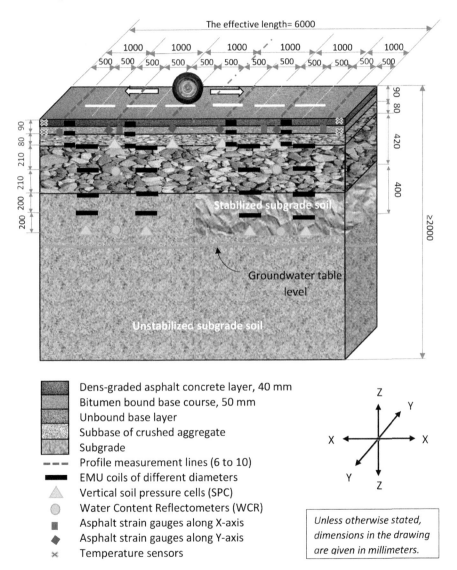

FIGURE 5.11 A recommended testbed constructed with surface stabilized subgrade soil, not to scale.

recommended testbed with unstabilized and shallow stabilized subgrade soil that should be able to withstand moisture damage more than the unstabilized one. Stabilized subgrade also helps to better withstand road damage caused by future autonomous convoy driving. In Figure 5.11, the road superstructure is divided into two halves. In one half, the subgrade is stabilized with a stabilization agent while the subgrade on the other half is left unstabilized.

The data to be collected from the accelerated loading testing for the testbed given in Figure 5.11 will help in evaluating the road performance under wet conditions and in numerically identifying the benefits of the stabilized section.

REFERENCES

AMA Anläggning. (2020): "Allmän material- och arbetsbeskrivning för anläggningsarbeten," Svensk Byggtjänst, ISBN 9789179170172. https://byggtjanst.se/bokhandel/ama/ama-anlaggning/ama-anlaggning/ama-anlaggning-20-bok?gclid=EAIaIQobChMI79zNh-Cn-QIVRoXVCh1PKw3pEAQYASABEgIzqvD_BwE

CFS. (2016): "Fabco stormwater solutions," California Filtration Specialists. https://california filtrationspecialists.com/products/cartridge-based-filtration/fabco-stormwater-solutions/

Climate Central. (2015): "Across U.S., Heaviest downpours on the rise," Across U.S., Heaviest Downpours On The Rise, Climate Central. https://www.climatecentral.org/news/across-us-heaviest-downpours-on-the-rise-18989

Franzén, G., Lindh, P., Åhnberg, H. and Erlingsson, S. (2012): "Terrasstabilisering-Kunskapsdokument," VTI rapport 747, Linköping, Sweden.

Global Street Design Guide. (2016): *National Association of City Transportation Officials,* NACTOs, New York, ISBN: 978-1-61091-494-9.

Hellman, F. (2017): "Accelererad provning av dränerande markstensytor med HVS – utrustning," Vinnova rapport för utmaningsdriven innovation – Hållbara attraktiva städer, Diarienummer: 2012–01271. http://klimatsakradstad.se/media/2017/11/HVS-rapport-Hellman-m-bilagor.pdf

Kåhre, R. (2020): "Uppvärmda vägar i stadsmiljö jämfört med traditionell snöröjning," Examensarbete för Byggingenjörsprogrammet på Mittuniversitetet, Östersund, Sweden. www.diva-portal.org/smash/get/diva2:1475720/FULLTEXT01.pdf

Kendon, M., McCarthy, M., Jevrejeva, S., Matthews, A., Sparks, T. and Garforth, J. (2021): "State of the UK Climate 2020," *International Journal of Climatology, The Royal Meteorological Society Journal of Climate Sience,* a special issue edited by Collins, B. and Aguilar, E., 41 (S). https://doi.org/10.1002/joc.7285

Selezneva, O.I., Jiang, Y.J., Larson, G. and Puzin, T. (2008): "Long term pavement performance computed parameter: Frost penetration," Publication No. FHWA-HRT-08–057, Research, Development, and Technology, Turner-Fairbank Highway Research Center, McLean, USA. www.fhwa.dot.gov/publications/research/infrastructure/pavements/ltpp/08057/08057.pdf

Sohrweide, Tom. (2018): "Driverless vehicles set to change the way we design our roadways?" SEH, published July 25, 2018. www.sehinc.com/news/future-what-do-driverless-cars-mean-road-design

SS-EN 13242. (2003): "Ballast för obundna och hydrauliskt bundna material till väg- och anläggnings byggande," svensk standard SIS, Stockholm.

State of the Global Climate. (2020): "World meteorological organization," Professional report. doc_num.php (wmo.int)

Sundberg, Jan. (2014): "Halkfria vägar – Förstudie -Solvärme och värmelagring för miljöanpassad halkbekämpning," Trafikverket rapport, Dokumentdatum: 2012–05–25,
Ärendenummer: 4805 Version: 1.0, Publikationsnummer: 2014:120, ISBN 978-91-7467-
646-4, Borlänge.

Sundberg, Jan and Lidén, Peter. (2014): "RAPPORT – Halkfria vägar – Etapp 2 Energi-
och systemanalys med kostnader Solvärme och värmelagring för miljöanpassad halkbekämpning," Trafikverket rapport, Dokumentdatum: 2014–10–21, Ärendenummer:
4805, Version: 1.0, Publikationsnummer: 2014:121, ISBN 978-91-7467-647-1, Borlänge.

Svensk Markbetong handbook. (2019): "Fördröjning av dagvatten med dränerande
markstensbeläggning- Projektering, Utförande samt Drift och Underhåll av mul-
tifunktionella Gaturum," Svensk Markbetong, ISBN 978-91-519-3477-8. www.
svenskmarkbetong.se/media/nwyeo1zp/svensk_markbetong_handbok_dranerande_
konstruktioner_180x255_webb_200330.pdf

The Swedish Commission on Climate and Vulnerability. (2007): "Sweden facing climate
change – threats and opportunities," Final Report, SOU 2007:60, Stockholm.

Swedish National Knowledge Centre for Climate Change Adaptation at SMHI. (2021):
Precipitation | Swedish portal for climate change adaptation (klimatanpassning.se).

TRVK Väg. (2011): "Trafikverkets tekniska krav Vägkonstruktion," TRV 2011:072, TDOK
2011:264, Trafikverket rapport, Borlänge.

USGCRP – U.S. Global Change Research Program. (2017): "Climate science special report,"
Fourth National Climate Assessment, I. Wuebbles, D.J., D.W. Fahey, K.A. Hibbard, D.J.
Dokken, B.C. Stewart and T.K. Maycock, eds. https://science2017.globalchange.gov.
https://doi.org/10.7930/J0J964J6

Willway, T., Reeves, S. and Baldachin, L. (2008): "Maintaining pavements in a changing
climate," Transport Research Laboratory, Published by TSO (The Stationery Office),
London. ISBN 978-0-11-552983-2.

World Climate Research Programme (WCRP) Global Sea Level Budget Group. (2018):
"Global sea-level budget 1993 – present," *Earth System Science Data*, 10, 1551–1590.
https://doi.org/10.5194/essd-10-1551-2018

6 Reconfigurable streets

GENERAL STATEMENT

The concept of reconfigurable streets has been highlighted recently and focuses on flexible street design that allows parts of the streets to be easily removed and reinstalled for maintenance work – or even changing the function that the street serves.

According to NR2C (2008), the clearest solution for smart design of a street network, in addition to expanding the street network, is to create temporary capacity inside and/or outside the network. Therefore, the first requirement is smooth connections and bypasses between different networks. A fairly easy way to achieve extra capacity in the road system is to allow the use of hard shoulders during rush hours and in the case of accidents. This means that consideration should be given from the design stage of streets that a lane allocated to vehicles, for example, can easily and economically be changed to a pedestrian and cycle path in the future or vice versa. Sometimes reconfiguration of streets includes street patching, signing, pavement marking and rubber curb installation, as well as allowing additional space for retail and dining activities outdoors. Parts of reconfigurable streets should be easy to remove and reinstall for maintenance work.

From the street construction perspective, the reconfigurable street concept requires building materials that can be easily and economically installed and removed, considering the technical and environmental characteristics of the materials used.

EFFECT OF THE RECONFIGURABLE STREET CONCEPT ON THE STRUCTURAL DESIGN OF ROADS

Concrete block pavement (CBP) is gaining popularity in urban areas because access to the underground utilities is a key requirement in urban road paving. Due to the segmental nature of the paving block, the concrete block is easily removed, and the underlaying utilities can be accessed easily for maintenance works. CBP is preferred as the ideal choice in urban areas.

A prototype of a modular paving system (concrete block pavement, CBP) that can be adapted to make the streets reconfigurable, safer and more accessible to everyone, has been used in specific projects and is known as Removable Urban Pavements (RUP).

Removable urban pavements are pavements that can be opened and closed within just a few hours, using very light equipment in place, and are able to restore a street to its original look and functionality (de Larrard et al., 2013). These pavements can provide many functions and incorporate many types of service networks (water and stormwater, electricity, telecommunications, etc.).

The slabs need to be mechanically independent of each other to be easy to lift during maintenance work. As for the joints, a soft, cold and waterproof polymer

DOI: 10.1201/9781003280224-6

material that is easy to remove manually can be poured into the joints. As with some conventional modular pavements, the slabs are placed over a gravel bed to simplify positioning.

According to de Larrard et al. (2013), though the base course served a structural purpose, network operators expressed a preference for using easy-to-dig material. Therefore, a new material should be used, called Structural Excavatable Cement-Treated Material (SECTM). This material has been added to the French pavement design specifications. A three-dimensional FEM calculation should be provided to evaluate the stresses caused by traffic in both the slabs and the base course. The allowable stresses can be determined according to the classical French pavement design fatigue approach. As a final step, the minimum slab and base thickness can be calculated. Based on the materials used by de Larrard et al. (2013), these parameters result in a 200 mm thickness for the traffic load and location conditions in a specific project, except around the edges where an additional thickness of 10 mm has been added to ensure good contact between the slab and the gravel bed. These calculations led to a total thickness of 600 mm for the base course, which consists of two layers of 300 mm of SECTM.

There are many types of similar roads commonly referred to as removable roads, portable roads and matting systems on the market today, and they serve the same purpose as RUP. For more details see Kuttah (2019b).

The concept of using removable and reconfigurable roads/streets is especially interesting from a road design perspective. Removable roads differ from traditional structural road design in making possible flexible lanes that can be changed from serving as walking and cycling paths to car lanes in the future and vice versa. Therefore, it is important to evaluate the long-term performance and suitability of many available removable road types, which requires further investigation with simulated long-term traffic loads at different climatic conditions.

If the technology of reconfigurable roads/streets is to be adopted in Sweden or other countries, proper studies should be carried out to bring new materials and construction methods into the local design standards. Afterwards, design concepts and manuals should be developed in which the thickness of slabs and base layers should be determined considering long-term traffic performance before wider use.

RECOMMENDATIONS FOR RECONFIGURABLE STREET TESTBEDS

Figure 6.1 shows a proposed testbed with RUP for testing with accelerated loading testing facilities. The results should then be compared with traditional road sections exposed to similar simulated testing conditions. The economic advantage of using RUP or other type of removable roads/streets should also be evaluated and compared with the costs of traditional paving systems.

Another alternative to making the current street design reconfigurable is to build the lanes for vehicles, pedestrian and cycle paths with the same subbase layers in terms of material and thickness, while the base layer is made thicker, and the asphalt layer is made thinner for pedestrian and cycle paths. Later, when pedestrian and cycle paths are opened to vehicles, the surface layer can be milled away together with the upper part of the base layer, and thicker asphalt can be laid to meet the requirements of driving vehicles lanes.

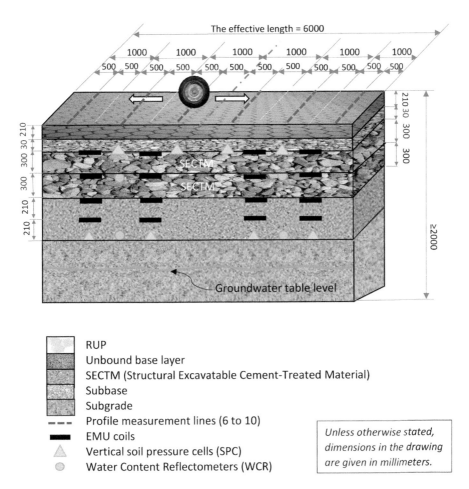

FIGURE 6.1 A recommended testbed constructed with removable urban pavements, not to scale.

REFERENCES

de Larrard, F., Sedran, T. and Balay, J. (2013): "Removable urban pavements: An innovative, sustainable technology," *International Journal of Pavement Engineering*, 14 (1), 1–11.

Kuttah, D. (2019b): "An overview on portable roads and airfields: Using of matting systems as temporary and semi-permanent roads and airfields," VTI notat; 15A-2019, Linköping, Sweden.

NR2C. (2008): "New road construction concepts -Towards reliable, green, safe & smart and human infrastructure in Europe," Final report by FEHRL, supported through the Sixth Framework Programme of the European Union. https://ec.europa.eu/transport/road_safety/sites/roadsafety/files/pdf/projects_sources/nr2c_final_report.pdf

7 Multifunctional road proposals

GENERAL STATEMENT

As discussed previously, it is important to test pavement design models to cope with the impacts of many future challenges, like urbanization, digitalization, electrification and climate change.

In this context, the functions of the testbeds proposed in this book have been summarized as given in Table 7.1 together with the recommended accelerated pavement testing (APT) test options and related observations.

PROPOSALS FOR POSSIBLE COMBINATION OF SMART SOLUTIONS DURING TESTING

It may be possible to verify a testbed that meets more than one challenge or adaptation at the same time and during the same HVS trial. Each testbed given in Table 7.2 has been fortified by several fixtures that allow the road section to withstand several challenges. This may be relevant in the case of inadequate testing funds and when it is not necessary to identify the influence of each challenge separately. If this is the case, the testbeds given in Table 7.2 are recommended to be tested using accelerated testing facilities (e.g., HVS facility).

Keep in mind that testing many solutions in one testbed can mix the individual effects in terms of double impact on the measured parameters. The possible risk that may arise for the selected test combination is discussed in Table 7.2.

Choosing the test priority for the road structures given in Table 7.2 depends on the current needs and time-schedules for adopting and implementing such technologies. Climate change has already begun, and only minor evaluation of road adaptation methods has been carried out so far; the infraculvert has already been used in Sweden, but before wider use, long-term evaluation in terms of durability under accumulated traffic loading and climate conditions should be carried out. Conductive and inductive technologies for charging electric vehicles have also been used in Sweden, but little data is yet available on their long-term durability under accumulated heavy traffic and difficult weather conditions.

Note that, in all the trials illustrated in Table 7.2, it is preferable to carry out the accelerated loading testing under different controlled groundwater levels to determine the change of groundwater levels on the measured parameters.

DOI: 10.1201/9781003280224-7

TABLE 7.1

Summary of the Proposed Testbeds and the Recommended APT Tests

Future challenge		The main feature of the suggested road testbed	Test suggestion		Testbed to be instrumented and tested	Notes
Main category	Purpose		HVS test at a fixed testing facility	In-situ test with a mobile HVS		
Urbanization	Reduce future network maintenance costs	Usage of underground infraculvert technologies		✓	Figure 2.3	Paver block surface
				✓	Figure 2.4	Paver block surface
				✓	Figure 2.5	Asphalt surface
				✓	Figure 2.6	Asphalt surface
Digitalization	To withstand platooning and channelized traffic caused by AVs, which may increase rut depths	Reinforcement under the wheel with concrete or steel mesh strip	✓		Figure 3.3	Performing HVS tests without lateral wander to simulate platooning and channelized traffic
		Constructing a thicker road structure	✓		Figure 3.4	
		Stabilization with asphalt reinforcement fibers	✓		Figure 3.5	
Electrification	To allow dynamic charging of electric vehicles	Electric rails for conductive charging / Inductive charging	✓		Figure 4.5	----------
Climate change	To control and reduce flooding	Usage of reservoir concept with porous asphalt surfacing	✓		Figure 5.4	----------
		Usage of storage tanks system	✓		Figure 5.6	----------
		Usage of Smart Trap	✓		Figure 5.7	----------
	Ice/snow reduction	Usage of pipes heating system	✓		Figure 5.10	----------
	Strengthen the road structure to withstand moisture damage	Stabilizing the upper 40 cm of the subgrade soil	✓		Figure 5.11	----------
Reconfigurable streets	Flexible design	Usage of removable urban pavements	✓		Figure 6.1	----------

TABLE 7.2

Proposals for Possible Combinations of Smart Solutions During Testing of Multifunctional Roads and Streets

Testbed ID	Suggestions for multifunctional roads testbeds that meet different future challenges					Combination features/Notes	Possible risk in judging each individual action/ technology due to the suggested testing combinations
	Urbanization	Digitalization	Electrification	Climate change	Reconfigurable streets		
1	Usage of underground infraculvert technology shown in Figures 2.5 or 2.6	Testing the asphaltic surface of type ABT 16 with a concrete or steel mesh strip reinforcement to withstand the simulated platooning and channelized traffic effect of AVs (Figure 3.3) + Performing HVS tests without lateral wheel wander				Combining the underground part of the testbed design given in Figures 2.5 or 2.6 (the infraculvert) with the superstructure given in Figure 3.3	The concrete or steel strip may decrease the actual HVS loading expected on the tested infraculvert
2	Usage of underground infraculvert technology shown in Figures 2.5 or 2.6	Stabilizing the wearing course with asphalt reinforcement fibers to withstand the simulated platooning and channelized traffic effect of AVs (Figure 3.5) +Performing HVS test without lateral wheel wander				Combining the underground part of the testbed design given in Figures 2.5 or 2.6 (the infraculvert) with the superstructure given in Figure 3.5	The reinforced asphalt may decrease the actual HVS loading expected on the tested infraculvert
3	Usage of underground infraculvert technology shown in Figures 2.5 or 2.6	Performing HVS tests without lateral wheel wander	Testing of a trial section with both conductive and inductive electrical vehicle charging technologies (Figure 4.5)			Combining the underground part of the testbed design given in Figures 2.5 or 2.6 (the infraculvert) with the superstructure given in Figure 4.5	There is a little risk that the charging rails and inductive slabs may reduce the actual HVS loading expected on the tested infraculvert. Also, in this combination only part of the digitalization (automation) requirement has been considered.

(Continued)

TABLE 7.2
(Continued)

Testbed ID	Suggestions for multifunctional roads testbeds that meet different future challenges					Combination features/Notes	Possible risk in judging each individual action/technology due to the suggested testing combinations
	Urbanization	Digitalization	Electrification	Climate change	Reconfigurable streets		
4	Usage of underground infraculvert technology shown in Figures 2.5 and 2.6	Performing HVS tests without lateral wheel wander		Usage of porous asphalt surfacing and the reservoir concept for stormwater storage (Figure 5.4)		Combining the underground part of the testbed design given in Figures 2.5 or 2.6 (the infraculvert) with the superstructure given in Figure 5.4	Not risky / In this combination only part of the digitalization (automation) requirement has been considered.
5	Usage of underground infraculvert technology shown in Figures 2.5 and 2.6	Performing HVS tests without lateral wheel wander		Usage of storage tank system (Figure 5.6) to control flooding and excess water		Testing 3 m of the underground part of the testbed design given in Figure 2.5 or 2.6 with the other 3 m of the storage system given in Figure 5.6	Not risky / In this combination only part of the digitalization (automation) requirement has been considered.
6		Performing HVS tests without lateral wheel wander		Usage of a smart trap system (Figure 5.7) and porous asphalt (Figure 5.4) to control flooding and excess water		Combining the underground part of the testbed design given in Figure 5.7 with the superstructure given in Figure 5.4	Not risky / In this combination only part of the digitalization (automation) requirement has been considered.
7		Performing HVS tests without lateral wheel wander		Usage of a pipes worming system to reduce ice/snow accumulation (Figure 5.10)		The same as in Figure 5.10 but considering only HVS test without lateral wheel wander	Not risky / In this combination only part of the digitalization (automation) requirement has been considered.

Testbed ID	Suggestions for multifunctional roads testbeds that meet different future challenges					Combination features/Notes	Possible risk in judging each individual action/ technology due to the suggested testing combinations
	Urbanization	Digitalization	Electrification	Climate change	Reconfigurable streets		
8		Performing HVS tests without lateral wheel wander		Usage of a pipes worming system to reduce ice/snow accumulation (Figure 5.10) + Usage of porous asphalt surfacing and the reservoir concept for stormwater storage (Figure 5.4)		The underground and the superstructure of the testbed are the same as the one given in Figure 5.4 using porous asphalt, but add the pipes worming system given in Figure 5.10	There is a little bit of risk since the porous asphalt may result in higher HVS loading on the piping system as compared to the traditional ABT 16 asphalt surfaces. Also, in this combination only part of the digitalization (automation) requirement has been considered.
9		Strengthening of the road superstructure by increasing its thickness to withstand the AV effect (Figure 3.4) + Performing HVS tests without lateral wheel wander	Testing of a trial section with both conductive and inductive electrical vehicle charging technology (Figure 4.5)	Strengthening of the road superstructure by increasing its depth will also increase the resistance of the road structure against moisture damage (Figure 3.4)		Combining the testing of the asphaltic layer with conductive and inductive electrical vehicle charging technology given in Figure 4.5 with the thicker superstructure given in Figure 3.4	Not risky
10	Usage of underground infraculvert technology with asphalt (Figures 2.5 or 2.6)	Stabilizing the upper 40 cm of the subgrade layer (for one half of the testbed) with a stabilizing agent to increase its resistance (Figure 5.11) will also improve overall performance and hence better adaptation to AVs (Figure 5.11) + Performing HVS tests without lateral wheel wander		Stabilizing the upper 40 cm of the subgrade layer (for one half of the testbed) with a stabilizing agent to withstand better the moisture damage (Figure 5.11)		Combining the testbed given in Figure 5.11 and install the infraculvert as given in Figures 2.5 or 2.6	There is a little risk that the stabilized subgrade soil may reduce the effect of actual HVS loading expected on the tested infraculvert as compared to using unstabilized subgrade.

(Continued)

TABLE 7.2
(Continued)

Testbed ID	Suggestions for multifunctional roads testbeds that meet different future challenges					Combination features/Notes	Possible risk in judging each individual action/technology due to the suggested testing combinations
	Urbanization	Digitalization	Electrification	Climate change	Reconfigurable streets		
11	Usage of underground infraculvert technology with paver blocks (Figures 2.3 or 2.4)	Stabilizing the upper 40 cm of the subgrade layer (for one half of the testbed) with a stabilizing agent to increase its resistant (Figure 5.11) will also improve the overall performance and hence better adaptation to AVs (Figure 5.11) + Performing HVS test without lateral wheel wander		Stabilizing the upper 40 cm of the subgrade layer (for one half of the testbed) with a stabilizing agent to withstand better the moisture damage (Figure 5.11)		Combining the test bed given in Figure 5.11 and installing the infraculvert as given in Figures 2.3 or 2.4	There is a little risk that the stabilized subgrade soil may reduce the effect of actual HVS loading expected on the tested infraculvert as compared to using unstabilized subgrade.
12	Usage of underground infraculvert technology with paver blocks (Figures 2.3 or 2.4)	Adopting the paver blocks superstructure given in Figure 5.3 with thicker subbase layer, which will help to better withstand platooning and channelized traffic of AVs + Performing HVS tests without lateral wheel wander		Adopting the paver blocks superstructure given in Figure 5.3 with thicker subbase layer which will work as an additional reservoir for the stormwater		Combining the test bed given in Figure 5.3 and installing the infraculvert as given in Figure 2.3 or 2.4	There is a little risk that the thicker subbase may reduce the effect of actual HVS loading expected on the tested infraculvert as compared to traditional subbase thickness.

Testbed ID	Suggestions for multifunctional roads testbeds that meet different future challenges					Combination features/Notes	Possible risk in judging each individual action/ technology due to the suggested testing combinations
	Urbanization	Digitalization	Electrification	Climate change	Reconfigurable streets		
13		Stabilizing the upper 40 cm of the subgrade layer (for one half of the testbed) with a stabilizing agent to increase its resistance (Figure 5.11) will also improve the overall performance and hence better adaptation to AVs (Figure 5.11) + Performing HVS tests without lateral wheel wander		Usage of the pipes worming system to reduce ice/snow accumulation (Figure 5.10) + Stabilizing the upper 40 cm of the subgrade layer (for one half of the testbed) with a stabilizing agent to better withstand moisture damage (Figure 5.11)		The test bed is almost like the one given in Figure 5.10 except that for one half of the testbed, the upper 40 cm of the subgrade soil will be stabilized with a stabilizing agent to better withstand the moisture damage	There is a little risk that the piping system may reduce the effect of actual HVS loading expected on the tested unstabilized half of the testbed.
14	Usage of underground infraculvert technology with paver blocks (Figures 2.3 or 2.4)	Performing HVS tests without lateral wheel wander			Usage of removable urban pavement blocks as shown in Figure 6.1	Combining the flexible street design technology given in Figure 6.1 and installing the infraculvert as shown in Figure 2.3 or 2.4	There is a little risk that the structural design given in Figure 6.1 may reduce the effect of actual HVS loading expected on the tested infraculvert as compared to using traditional materials. Also, in this combination only part of the digitalization (automation) requirement has been considered.

PROPOSALS FOR POSSIBLE COMBINATION OF SMART SOLUTIONS IN REAL APPLICATIONS

Due to the importance of designing flexible streets, it has become necessary to highlight the possibility of using the new, proposed technologies on streets that can be reconfigured in the future. Nowadays many lanes serving as car lanes may be reconfigured to bike/pedestrian lanes in the future and vice versa. The lanes designed to be pedestrian/cycle paths cannot be directly reconfigured into car lanes without redesign or adaptation of the road structure; otherwise it will be undersized and cannot withstand the traffic loading. On the other hand, the lanes designed to be car lanes can be directly reconfigured to pedestrian/cycle lanes without redesign or adaptation of the road structure – except if the car lane has electric rails for conductive charging of electric vehicles. In this case, the charging rail should be removed, and the rail groove buried so the car lane can safely be used as a bike/pedestrian lane. Also, caution should be taken when using a car lane with drained paver blocks or removable urban pavement to serve as a bike/pedestrian lane. In such case, the suitability of the used surface material to meet the requirements for accessibility and wheelchairs should be checked.

Regarding the streets and roads that should be ready to withstand several future challenges at the same time, Table 7.3 illustrates models and proposals for a possible combination of several technical solutions for real applications.

Another important question is whether the proposed technologies can be used only during new construction of roads and streets, or also for existing ones. The answer is that there are no obstacles from a technical perspective to use all the suggested techniques on existing roads. Even so, launching a technology during new construction tends to be cheaper than modifying existing roads, which may require partial or extensive reconstruction. So, it is more of an economic issue than a technical one. Note that for existing roads, the features installed deeper in the street/road construction will require deeper excavation work and thus the cost of traffic delay may be significantly higher. In adopting the proposed models given in this book for construction to withstand future challenges, the suitability of the materials and technologies from the environmental point of view should also be approved based on the current local national and/or international standards.

Note that when reconstructing existing streets in an inner-city environment – for instance, to be equipped with underground technologies at great depths, as in a few of the solutions suggested in this book – archaeological examination of historical layers may be required before excavating. In addition, congestion under the street from wires and large root systems should be considered during the reconstruction process.

TABLE 7.3
Proposals for Possible Combination of Smart Solutions in Real Application of Multifunctional Roads and Streets

Suggested combination	Urbanization	Digitalization				Electrification		Future challenges — Climate change					Reconfigurable streets
	Using of underground infraculvert	To withstand platooning and channelized traffic of AV — Using of a concrete or steel mesh strip under the wheels	Constructing thicker superstructure	using fiber stabilized asphalt	Speed control — Using active speed bumps	Electric rails	Inductive charging	Using of reservoir concept with porous asphalt	Using of Storage-Tank system	Using of Smart Trap	Using of pipes worming system	Stabilizing the upper 40 cm of the subgrade	Using of removable urban pavements
1	✓		✓	✓							✓	✓	
2			✓							✓		✓	✓
3	✓	✓	✓		✓							✓	
4			✓									✓	
5			✓	✓	✓	✓	✓	✓	✓			✓	
6	✓		✓	✓	✓						✓	✓	
7			✓	✓						✓	✓	✓	
8		✓	✓		✓			✓				✓	
9	✓		✓	✓		✓						✓	
10	✓		✓	✓	✓		✓					✓	

8 Conclusion – toward smart roads

Modern roads and street design can solve many challenging problems related to future urbanization, digitalization, electrification and climate change. Until now, few measures have been undertaken to deal with such challenges. These measures have influenced mainly the geometric design features of streets and roads and can be manifested by the use of electrical vehicle charging stations along the sides of the roadway, as well as usage of swales and rain gardens for stormwater management. The concepts of swales and rain gardens have been adopted recently in many countries as good solutions for managing stormwater. Although swales and rain gardens add a touch of green to the streets, in cases where space is lacking it will be difficult to use them.

To deal with these challenges in a more comprehensive manner, many modern smart solutions that influence the structural design of roads have been illustrated and discussed in this book. These measures can be considered flexible and practical to catch up with future challenges. One of the solutions is manifested in the use of underground infraculverts that can prevent future excavation of the roads for pipe maintenance and thus help keep the roads open during maintenance work. Furthermore, the use of electric vehicles will no longer require charging stations along the sides of the roadway as the new concept of dynamic charging of vehicles has been adopted. This concept will allow charging of vehicles while driving with the help of technology embedded in the road structure – namely, electric charging rails for conductive charging and copper coils and rubber systems for inductive (wireless) charging.

Future digitalization will also affect the configuration and even the structural design of the roads and streets to withstand autonomous vehicles. The use of autonomous and semi-autonomous trucks can lead to an even narrower use of lanes. The transition to autonomous vehicles assumes a concurrent reduction in lane width (about 25% reduction in freeway lane width) and an increase in roadway capacity (more than 50% by several industry estimates). For roads and streets with multifunctional use, narrower lanes allow more space to be dedicated to other activities.

The use of autonomous vehicles (AVs) will affect the structural design of roads due to the effects of more channelized traffic and convoy driving. This can lead to higher traffic volumes of more uniform vehicle types and loads that could bear on the pavement structure in ways it had not been designed to bear, accelerating the accumulation of pavement damage. Strengthening the road structure with a reinforced strip (concrete or steel mesh) placed under the wheel tracks of AVs will help in reducing the damage of the road. Alternatively, a thicker road construction or even strengthening the asphalt surface with asphalt-strengthening fibers will reduce ruts and cracks that can be developed dramatically after the road is opened to AVs.

DOI: 10.1201/9781003280224-8

New speed control technologies that fall under the digitalization theme are now in use and they are related to the need for street design that is more multifunctional and traffic safe. One such speed control technology is the installation of active speed bumps based on dynamic speed. Vehicles exceeding the speed limit activate a gap that is lowered a few centimeters into the roadway and give the driver a physical reminder of the speed violation. The system allows everyone who drives at the right speed to pass on a flat road and therefore does not pose a problem for bus drivers. Regarding adaptation of roads and streets to withstand climate change, there are also many technologies adopted recently by many actors and municipalities. An interesting solution from a structural design point of view is the usage of underground solutions to handle stormwater. These embedded technologies allow instant stormwater management and even water filtration within the road structure. One of these systems is the use of drained paver blocks or porous asphalt to allow infiltration of stormwater to the lower road layers, which can be made thicker to provide good storage of stormwater. Note that the suitability of any type of drained paver blocks to meet the accessibility requirements for handicaps and wheelchairs should be checked before any wider use.

Moreover, the use of storage tanks and Smart Trap concepts will offer immediate management of large amounts of stormwater and prevent possible flooding of streets.

The other climate-related problem is moisture-related damage to road construction. The moisture susceptibility of subgrade soil can be reduced by shallow stabilization of the subgrade layer (e.g., by using modern environmentally friendly additives).

Regarding the problem of accumulated snow during winters, the use of underground pipes heating systems will increase the passability of roads and make them accessible and open even during heavy snowfall. Usage of heating systems via pipes will reduce the cost of traffic interruption du to snow plucking (no plucking machines or salt or sanding machines are needed anymore).

The concept of reconfigurable streets has also recently been highlighted as a flexible street design concept. From a technical design perspective, reconfigurable and removable roads/street concepts require the use of building materials that can be easily and economically installed and removed. The concept of reconfigurable streets is also important to enable possible adaptation of an existing bike path to be used for vehicles driving during acute and emergency situations. Prototypes of modular paving systems that can adapt to people's changing needs, making the streets reconfigurable, safer, and more accessible for everyone, have been used in individual projects. If this technology is to be used more widely, detailed studies should be carried out to incorporate the materials and construction methods according to local design standards. Then, for each specific project, the thicknesses of slabs and base layers should be verified and determined. The environmental aspect of the new materials should also be assessed.

Another important question is whether the proposed technologies can be used only during new construction of roads and streets, or also for existing ones. The answer is that there are no obstacles from a technical perspective to use all the suggested techniques on existing roads. However, comprehensive studies are required between transport demands and infrastructure demands and between the economic operation and maintenance cost of the new technologies.

Certainly, all the proposed measures in this book have different impacts on the structural design of roads and streets. So, the performance of each of the proposed designs needs to be evaluated at full scale under long-term traffic loads and different climatic conditions, preferably using accelerated loading testing, to ensure their durability and sustainability before wider use.

Index